Synthesis Lectures on Mathematics & Statistics

Series Editor

Steven G. Krantz, Department of Mathematics, Washington University, Saint Louis, MO, USA

This series includes titles in applied mathematics and statistics for cross-disciplinary STEM professionals, educators, researchers, and students. The series focuses on new and traditional techniques to develop mathematical knowledge and skills, an understanding of core mathematical reasoning, and the ability to utilize data in specific applications.

Apostolos Syropoulos

Fuzzy Mathematics

A Fundamental Introduction

 Springer

Apostolos Syropoulos
Xanthi, Greece

ISSN 1938-1743   ISSN 1938-1751 (electronic)
Synthesis Lectures on Mathematics & Statistics
ISBN 978-3-031-73833-3   ISBN 978-3-031-73834-0 (eBook)
https://doi.org/10.1007/978-3-031-73834-0

This Springer imprint is published by the registered company Springer Nature Switzerland AG
The registered company address is: Gewerbestrasse 11, 6330 Cham, Switzerland

If disposing of this product, please recycle the paper.

Dedicated
to my son Demetrios-Georgios
and Linda

Preface

In 2022, I received an invitation to contribute a chapter for a book about the use of neutrosophic sets (an extension of fuzzy sets) in education. I asked some of my colleagues whether they would like to work on a small-scale project on the introduction of fuzzy sets to secondary education students. They liked the idea and we started working on the project. Two of my colleagues taught the material that I prepared to two different classes consisting of 25 students. The fourth member, who is the principal of the school where we work, provided all the necessary help and assistance required. The result of our endeavor was documented in a chapter that was published as a chapter of a book.[1]

When the project was concluded, I thought it would be a nice idea to *transform* the material I had prepared for this mini course into a little book. At that time, I also noticed that there was no simple but complete introduction to fuzzy mathematics. Thus, the idea seemed reasonable and I am really glad that Springer agreed with me. The book that you hold in your hands (or view on a computer screen, a tablet, or a smartphone) is the result of this endeavor. Our mini project revealed certain things (e.g., what is the minimum mathematical background to follow such a course or read a book based on it) and I took under serious consideration all of them. Consequently, the book has been designed for anyone who has a very basic background in mathematics and wants to learn what are these weird mathematical objects that are called *fuzzy sets*. Anyone who wants to learn about fuzzy sets needs to know what is a set and what are the properties and operations between sets. Thus, this book introduces all these notions and ideas. In different words, it is a self-contained introduction to the basics of fuzzy mathematics. However, fuzzy sets are a mathematical model of *vagueness* and Chap. 1 of the book is an introduction to vagueness and related ideas. This chapter presents the various concepts and ideas from a philosophical and a practical point of view. Since fuzzy sets are sets, it is necessary to

---

[1] Syropoulos, A., Giakati, I., Prountzos, I., & Tatsiou, E. (2023). *Introducing Vagueness in the Mathematical Curriculum of Secondary Education: Experience in Greece.* In S. Broumi (Ed.), *Handbook of Research on the Applications of Neutrosophic Sets Theory and Their Extensions in Education* (pp. 205–214). IGI Global. https://doi.org/10.4018/978-1-6684-7836-3.ch010.

have a basic understanding of sets and the operations between them. This is the subject of Chap. 2. Chapter 3 introduces fuzzy sets, their operations, their representations, and some of their most common extensions. Chapter 4 is about fuzzy numbers and their operations. Fuzzy numbers have found many applications and, naturally, most people working on practical problems using fuzzy sets, use fuzzy numbers. Certainly, this book could have more about fuzzy sets and their mathematics, but this was not my initial intention. The reader who will read this book and get an understanding of fuzzy sets and their operations can continue studying fuzzy sets and their mathematics—there are many good books on the subject.

Xanthi, Greece   Apostolos Syropoulos
July 2024

Acknowledgments I would like to thank my colleagues Ilias Prountzos, Ioanna Yiakati, and Eleni Tatsiou for their excellent collaboration. I thank Susanne Filler for believing in this little project and Boopalan Renu for patiently waiting for the project to finish.

Contents

On Vagueness

Indeed, beginning with Frege and for a very long time, logicians
have been assuming that vagueness may bring logical disaster in its
wake.

—Achille C. Varzi
(Italian philosopher)

Vagueness, uncertainty, and imprecision are properties or qualities that people confuse and take one for the other. The fact that we use different words for them, is not a proof that they are different. On could argue that these words are just synonyms. However, here I will (try to) explain the notion of vagueness and how it differs from uncertainty and imprecision.

1.1   What Is Vagueness?

There are many words like the words tall, beautiful, fat, etc., that have one thing in common: They do not convey a precisely defined meaning. I am sure the reader will agree with me that there is no general consensus on who is a tall person, who is a beautiful person, etc. Let me give an example to make things more clear. Consider a person whose height is 170 cm (5ft 6in). Is this a tall person? If the person is a male, we are not really sure; but if the person is a female, then it is definitely a tall person, particularly if she is from some Mediterranean country. On the other hand, although we have learned that the opposite of, say, beautiful is ugly, still this is not really true: There are millions of people out there that are not beautiful but they are not ugly either. They are just ordinary people like me.[1]

Colors are something quite interesting. First of all, people do not sense colors the same way. This is why some really weird things happen. I am really sure that you have experienced

---

[1] This section is based on [62].

the following scenario: You are standing in front of a very beautiful and colorful landscape and you decide to take several pictures of it with your digital camera or your smartphone. Later on, you check these pictures on your computer or on your smartphone and you realize that the colors are not those you expected! Why? One explanation could be that something is wrong with the device that took the pictures. I have already given an explanation of why this happens: Color is a subjective sensation. This means that we cannot directly measure or describe a color. To better understand this, consider the color "red," then a good question is: Which one of the following colors are more red or less red than the others?

Light Coral	Orange Red	Tart
Red	Scarlet	Vermilion
Crimson	Rufus	Maroon

Some people will claim that maroon is too red and at the same time they will claim that light coral is not red at all. Others may claim that red and scarlet are the same color (trust me they are not!). These and other similar remarks, show that we cannot agree on things that we thought that were clear on our minds! Let me note that there are color models—mechanisms by which we can describe the color formation process in a predictable way—to represent colors and so use them in all possible ways. Still the problem with these models is that not all colors can be described (the models are discrete while colors lie on a continuum) and the fact that there are many models means that there is no unique way to describe a color. Of course, these color models are quite useful as they allow people to describe some colors in a precise way. Finally, let me assure you that nothings is wrong with your digital camera or your smartphone!

In a nutshell, for color, height, and many attributes, it is difficult to definitely say that an object has a particular property. The boundaries of these attributes are fuzzy. This is why we cannot easily say whether something is light green or not. The boundaries depend on context, on our view of things, etc.

The Sorites Paradox, which was introduced by Eubulides of Miletus, is a typical example of an argument that demonstrates fuzzy boundaries. The term "sorites" ("σωρείτες") derives from the Greek word "soros" ("σωρός"), which means "heap," since the paradox was originally about the number of grains of wheat that make a heap. All agree that a single grain of wheat does not comprise a heap. The same applies for two grains of wheat. Or three. Or four…However, there is a point where the number of grains becomes large enough to be called a heap. Except there is no general agreement when this occurs (see Fig. 1.1). A similar paradox is the bald man paradox. Clearly, a man who has a full head of hair is not a bald person. If he loses one hair he will still have a full head of hair. The same applies if he

Fig. 1.1 A depiction of the
Sorites Paradox (drawing by
Ioannis Kontovos)

Fig. 1.2 Koula the dog!
Unfortunately, she passed away
on February 21, 2024 and made
me very sad that day

loses two, three, four, etc., hair. However, if he loses enough hair he will become bald but
there is no particular number of hair that will make this man a bald man.

In general, when we cannot agree whether someone or something has a particular property,
then we say that this property is vague. Some advocate that since we lack crucial information
that prevents us from properly categorizing a particular person or object, we see the persons
or objects vaguely. This is the *epistemic* view of vagueness. Others argue that the languages
we speak have deficiencies which prevent us from properly judging whether a shirt is blue
or not, or whether a person is tall or not. This is the *semantic* view of vagueness. However,
there is a third view of vagueness, according to which objects, persons, animals, plants, etc,
are really vague. This is the *ontic* view.

In order to understand this idea let us consider my dog Koula (see Fig. 1.2). She is
constantly losing hair, but at the same time new hair grows. So strictly speaking, Koula at
14:00 is not the same dog as Koula at 20:00, since during these six hours she may lose hair,
she may eat something or poop, etc. In the end Koula at 20:00 would be a slightly different
dog from Koula at 14:00. A skeptical reader may object, claiming that it is the same dog
based on the idea that the dog is essentially the same. By following this train of thought we
can conclude that an old and a young person are the same. In one sense they may be, but in
many senses they are not. Not to mention that people change in general.

Let us forget about dogs and their hair, and think instead about geometrical objects like cubes and spheres. Everyone who has been taught some school geometry knows the properties of these objects. For example, a cube has 6 faces, 12 edges, and 8 vertices, and all edges have exactly the same length. Now the question is: Are there cubes in the real world? In different words, are there such crisp (that is, non-vague) objects in the world we live in?

Perhaps surprisingly, the answer is that there are no cubes in the real world, but there are objects that are approximately cubes. This means that the edges have approximately the same length and that the faces are approximately parallel. More generally, there exist only approximations of pure abstract mathematical objects.

As was noted above, not everyone believes there are vague objects. In fact, in a short paper entitled "Can There Be Vague Objects?" [20] Michael Gareth Justin Evans "proved" that there are no vague objects. He is stating by assuming that an object a^2 is vague if there exists an another object b such that we cannot say whether a is identical to b. In different words, we cannot tell whether $a = b$ is true or false. Such statements are called indeterminate. A simplified version of Evans's proof, borrowed from [76], goes as follows:

(1) The statement $a = b$ is indeterminate.
(2) From the previous statement we can conclude that b is indeterminately equal to a.
(3) The statement $a = a$ is determinate.
(4) From the previous statement and (1) we conclude that a is not indeterminately equal to a.
(5) From (2) and (4), we conclude that a is not equal to b.

Conclusion (5) says that there are no vague objects! One basic objection to this argument is that it uses two truth values[3] to argue about something that requires definitely more truth values. However, even if one forgets this and attempts to understand the proof in a setting that admits only two truth values, then it is possible to show that Evans's proof holds while this does not excludes the existence of vague objects! For example, there is an *axiom* (i.e., a statement that is regarded as being established, accepted, or self-evidently true) in set *theory*[4]

---

[2] The symbol "a" stands for any object, that is, not a specific object, and in mathematics it is called a *variable*. In mathematics, variables are *names* (e.g., letters, words, special symbols) that usually stand for mathematical values (e.g., numbers, geometric shapes, etc.).

[3] Most people assume that any statement is either true or false. These two words or the numbers 0 and 1 that correspond to false and true, respectively, are called truth values. However, it is possible to have more truth values by assuming that 0 corresponds to (absolute) false, any natural number n corresponds to (absolute) true, and the natural numbers in between them to the additional truth values. Thus, a statement can be true to some degree.

[4] A theory is a plausible or scientifically acceptable general principle or body of principles offered to explain phenomena. Alternatively, a theory is the general or abstract principles of a body of fact, a science, or an art. In simple words, a theory is an idea or set of ideas that is intended to explain facts or events. Definitions from: "Theory." Merriam-Webster.com Dictionary, Merriam-Webster, https://www.merriam-webster.com/dictionary/theory. Accessed 7 Apr. 2024.

that is called *axiom of extension* (see Sect. 2.1). This axiom asserts that two sets are equal if they contain the same elements. Shunsuke Yatabe and Hiroyuki Inaoka [76] proposed that a is a vague object if and only if the axiom of extension does not hold for a. The interesting things is that the axiom of extension has no real importance for the formalization of mathematics. This means that we can have vague objects in a mathematical universe where Evans' proof holds. Of course, we could refute the proof by noting that it reasons about vague objects in a setting where only two truth values are admitted, but this a quite an elaborate task [14].

The discussion so far may you ask yourself: If vagueness is the norm, then why does science ignore this very important fact? A possible answer is that people have a deep respect for pure, abstract objects with exact properties, and everything in-between makes us feel uncomfortable. According to the following excerpt from "The Life of Marcellus," which is part of Plutarch's "The Parallel Lives," we see that Plato was against the study of curves that could be drawn only with the help of mechanical devices[5]:

> For the art of mechanics, now so celebrated and admired, was first originated by Eudoxus and Archytas, who embellished geometry with its subtleties, and gave to problems incapable of proof by word and diagram, a support derived from mechanical illustrations that were patent to the senses. [...] But Plato was incensed at this, and inveighed against them as corrupters and destroyers of the pure excellence of geometry, which thus turned her back upon the incorporeal things of abstract thought and descended to the things of sense, making use, moreover, of objects which required much mean and manual labor. For this reason mechanics was made entirely distinct from geometry, and being for a long time ignored by philosophers, came to be regarded as one of the military arts.

For Plato only curves that could be drawn using ruler and compass are pure and thus worth studying.

In the physical sciences, vagueness sometimes turns up in the most unexpected places, such as the world of computing. When engineers were building the first computers, they needed to find a way to represent information. Encoding dozens of symbols, or characters, is difficult, because one needs many distinct physical states to represent each character. However, it is possible to represent each character by a number through a "mapping," which simply means that we can construct a table where characters are mapped to consecutive integers: $a \rightarrow 1, b \rightarrow 2$, etc. Since the decimal numbering system uses ten digits to encode numbers, we would need ten distinct physical states to represent these ten digits. But the decimal numbering system is not the only one. The simplest is the binary system, which has only two digits, "0" and "1." Encoding two digits in an physical system is relatively easy, and one can exploit electric currents to represent these two digits. Thus the people who built early computers decided that when current travels along a wire, we should assume that the

---

[5] Text available from
https://penelope.uchicago.edu/Thayer/E/Roman/Texts/Plutarch/Lives/Marcellus*.html.

digit "1" has just traveled along the wire; similarly, when no current travels along a wire, we should assume that the digit "0" has just traveled along the wire.

But how do we detect if current travels along a wire or not? We measure the potential difference between two parts of the circuit. More specifically, if the potential difference at a given part of the circuit was, say, 3.5 Volts, then the computer pioneers assumed that the digit "1" had just passed through this part. If the potential at the same part of the circuit was, say, only 0.3 Volts, then they assumed that the digit "0" had just passed through it. The problem here is that the potential is not always exactly 3.5 Volts or 0.3 Volts. Voltage fluctuations happen, perhaps because there are loose or corroded connections either at base or on the power-lines, or because of bad weather, or extreme heat, and so on. Indeed, there are many things that can cause voltage fluctuations, and so the measurements are always approximate. In a nutshell, although computers are supposed to be precise devices, they operate on imprecise power sources and differences and thus they are vague by nature. Naturally, the computer pioneers wanted to be able to compute precise quantities, and so they did everything to get rid of the extra feature of the power being measured, its in-exactitude. Yet, as various more modern approaches to computing have revealed, it is possible to incorporate vagueness to get better results, with systems that operate more smoothly than their counterparts which completely ignore vagueness.

1.2   Vagueness, Uncertainty, Imprecision…

There are some notions that look similar to vagueness. For example, uncertainty looks surprisingly similar to vagueness but they are not the same. In different words, these notion have subtle differences but their differences are enough to characterize different situations or things. In order to understand the difference between vagueness and some similar notions, it would be nice to know when something that is uttered by someone is vague. A general definition of vague sentences would be quite helpful and the following one, which is an adaptation of the one given by Otávio Bueno and Mark Colyvan [12], does the trick:

Definition 1.2.1 A sentence is vague just in case it can be employed to generate a sorites argument.

The formulation of the original definition used the word *predicate* but in order to keep things simple I have opted to replace it with the word sentence. As expected, this definition is not precise but then it would be too much to ask for a precise definition of vagueness.

The term *ambiguity* is about sentences that have more than one meanings. A classical example of ambiguous statements are Pythia's oracular statements. It is known that Pythia, who was the priestess of the Temple of Apollo at Delphi, never gave a straight answer to people who asked for her advice (I suspect she did it to be always on the safe side…). For example, the following statement is what Pythia answered to the Athenians when asked how to defend against the Persians.

> A wall of wood alone shall be uncaptured, a boon to you and your children.

Other such statements are listed in the The Mythscapes blog.[6] It is not difficult to see that the previous statement cannot be used to generate a sorites paradox. More generally, we can safely assume that no ambiguous statements can generate a sorites paradoxes.

According to [32] *precision* is "[t]he closeness of agreement between independent test results obtained under stipulated conditions." Alternatively, we could say that precision is the degree to which independent measures agree with each other. Thus precision and imprecision are closely related to measurements. If vagueness is a fundamental property of this world, then clearly it is next to impossible to have precise measurements, therefore, one could say that imprecision is a consequence of vagueness, but, clearly, the two notions are different.

Generality is yet another word that is close to vagueness. General terms are words like "chair." There are many and different kinds of chairs, nevertheless, all the chairs that one may imagine are still chairs. Thus, a general term is one that incorporates the (few?) common characteristics of a group of similar objects. General terms are used to make up classes of objects.

It is not widely accepted that chance plays an important role in nature. There are people who believe that our universe is completely *deterministic* but there are others who assume that the universe is *nondeterministic*. In such a universe randomness has a central role to play. In a way, randomness is a guarantee that we have no way to say what will happen next. Naive probability theory [i.e., not the one formulated by Andrei Nikolaevich Kolmogorov [37] (Андрей Николаевич Колмогоров)] is an attempt to make predictions about future events. Naturally, these predictions depend on the assumptions we make. But then the real problem is to what extent these assumptions are meaningful. Certainly, I will not attempt to discuss what meaningful actually means…

One more term that is remotely related to vagueness is *uncertainty*. This notion is related to a situation where the consequences, the extent of circumstances, the conditions, or the events are unpredictable. In quantum mechanics the uncertainty principle is, roughly, about our inability to accurately measure at the same time the location and the momentum of a subatomic particle (i.e., an electron). Figure 1.3 shows a conventional picture of an atom. Here electrons take positions that look quite predictable. Naturally, the picture is dead wrong but textbooks still use drawings like these to explain a number of things. In general, we know that an electron is a wave and not a "solid" particle but for the sake of the argument we can assume that it is a sphere that vibrates extremely fast. So an electron would be a vague object if it would be in all these positions at the same time but the uncertainty springs from the fact

---

[6] See:
https://mythscapes.wordpress.com/2013/02/09/top-5-of-pythias-oracular-statements/.

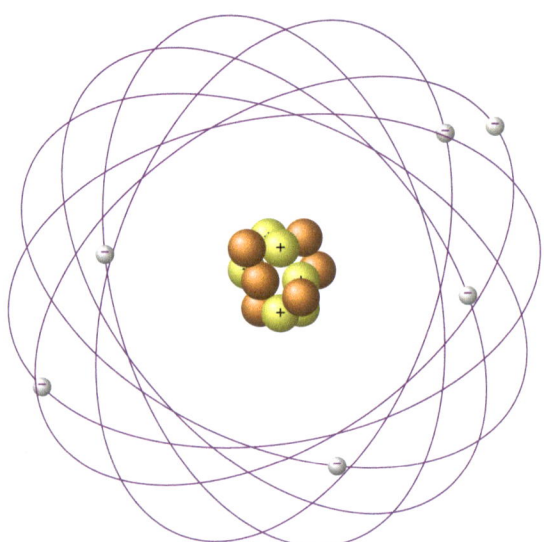

Fig. 1.3 A drawing depicting an atom

Fig. 1.4 Lotfi Aliasker Zadeh

that it moves so fast that we cannot spot its place. In reality, an electron as a wave is in all places at the same time. Moreover, one can say that an electron is a cloud, so an electron is actually a vague object. Interestingly there is a deep connection between uncertainty and vagueness through possibility theory.

When Lotfi Aliasker Zadeh (1921–2017) (Fig. 1.4), the founder of fuzzy mathematics, introduced fuzzy sets [77] he justified his work by using examples of vague concepts like "the class of beautiful women," and "the class of tall men." Fuzzy sets are the cornerstone of fuzzy mathematics (i.e., a form of vague mathematics) just like *sets* are considered by most mathematicians as the cornerstone of (ordinary) mathematics. For now, let us just say that Zadeh's fuzzy sets are a tool that can be used to classify and reason about vague objects. Unfortunately, there was a *confusion* in the community of people working and/or using fuzzy mathematics. In fact, people working on fuzzy mathematics got divided into two groups: those who believed that fuzzy mathematics are mathematics of vagueness and those who believed that fuzziness and vagueness are two different things. In fact, the second group advocated the idea that probability and fuzziness are two facets of uncertainty. Members of the second group argued that there is some sort of misunderstanding between the two communities:

> One of the reasons for the misunderstanding between fuzzy sets and the philosophy of vagueness may lie in the fact that Zadeh was trained in engineering mathematics, not in the area of philosophy. [17]

This argument is completely silly as there is no engineering mathematics but just mathematics. Also, one cannot compare (presumably applied) mathematics with philosophy as one cannot compare apples with oranges. Nevertheless, even philosophers employ mathematics to develop and to formalize their own theories and ideas. What is really interesting is that some philosophers advocated the idea of truth degrees something that is central to the theory of fuzzy sets, in particular, and fuzzy mathematics, in general (e.g., see [55]).

1.3   Measuring Vagueness

Typically a non-vague event is either true or false. Similarly, we can say that is it either true or false that a specific object has a particular non-vague property. Sometimes, we use the numbers 0 and 1 to represent the *values* true and false, respectively. Let me present a few examples, that will make clear what it means that a statement is true of false. If I say that in 2023 Manchester City won the Champions League title, this is a true statement. Also, if I say that in 2023 Inter Milan won the Champions League title, this is a false statement. However, if say that AEK Athens will win the the Champions League title in 2030, this is a statement that is neither true nor false. In fact, Aristotle, the father of Logic (i.e., the study of the principles of correct reasoning), was the thinker who spoke about statements that are neither true nor false. Although the third statement is not a vague, it gives an idea of how we could *measure* vague objects and events. In fact, there are at lest four different approaches to this problem but I will provide a general discussion of one approach. Readers who might be interested in reading more about the other approaches should consult the literature (e.g., see [59, 65]).

   One important problem that we need to tackle is whether we can objectively measure vagueness just like we measure the weight or the height of people. Although the weight of a person may depend on the context (e.g., when measuring the weight of person on Earth and on the Moon, the result of the measurement is different), still there are ways to approximately measure the weight of any person. I hope that the examples in Sect. 1.1 have made it clear that in the most general case two different people will not agree that some object has some quality. Usually, we use *linguistic modifiers*, that is, words like "many," "little," "few," etc., to say that a blouse is blue or yellow. For example, the expression "this blouse is very blue" indicates that quite possible most people will agree that the given blouse is blue. But in science we cannot rely on linguistic modifiers; we need something more...precise.

Nicholas Jeremy Josef Smith [55] has argued that the degree of truth, t, of a given statement could be either any integer number from a given range or a number that is greater than or equal to zero and less than or equal to one ($0 \leq t \leq 1$). Although both approaches have their pros and cons, still the vast majority of authors believe that the second approach is better. However, there is a "problem" with this approach—there are numbers in this range that cannot be "computed" with our current computing devices. For example, if we say that the degree of truth of a statement is the number $\pi/8$,[7] then we cannot have a decimal representation of this number. The real problem here is that $\pi/8$ has infinite decimal digits so we cannot have a conventional representation of this number. On the other hand, there are numbers whose digits are solutions...to insoluble problem, that is, problems that we do not know how to solve. Yet, I do not think this is a real problem. First, I am more than sure that no one will consider as truth degree a number whose digits correspond to the solution of insoluble problems—it simply makes no sense! Second, the number $\pi/8$ is not non-computable as we do know how to compute the digits of the number π. We assume we can live happily with an approximation of this number; after all people are sending spacecrafts to other planets, have computed their trajectories with approximations of the number π, and nothing goes wrong. In addition, we can assume that there are geometric or other mathematical constructs that faithfully represent numbers like π.

Now that we have chosen a way to assign a truth degree to statements, we need to come back to the previous question: Is there an objective way to assign degrees of truth? Unfortunately, the answer is No. The reason is that each of us has his/her own way to view things and so we rarely agree on the degree of vagueness. And this is why people have proposed some more complicated ways to assign truth degrees and I will come back on this matter in Sect. 3.4.

In conclusion, assigning a truth degree to a statement or an event is a subjective process but people may agree up to some degree.

---

[7] The symbol π represents the ratio of the circumference of a circle to its diameter and it is approximately equal to 3.14159.

Fig. 1.5 Is a circle after removing one point from it still a circle?

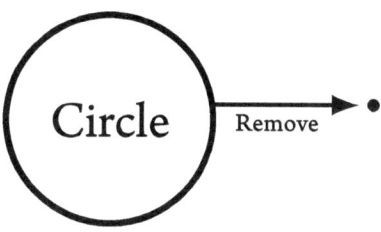

1.4 "Continuous" Sorites Paradox

In a recent post in his X account (formerly known as Twitter), Clifford Alan Pickover asked the following question:

> A mathematician emerges from a cave, hands you the slip of paper below, and says "Does a circle with one point removed have fewer points than the original circle? Is it still a circle?" What is your response?

Figure 1.5 depicts the problem. Mathematically speaking, a circle has infinitely many points and if you remove one single point, then what is left consists of infinitely many points. However, the circle and the circle-with-point-removed are very different objects. In fact, the later is, as we say in mathematics, *homeomorphic* to the real line. Technically speaking, a circle is what we call a *compact set* but after removing a single point it is not a *closed set*. The reason is that a *compact set* is something that is both *closed* and *bounded*, therefore, the circle-with-point-removed is not *compact*. So the circle is a *compact* object and the circle-with-point-removed is not a *compact* object, therefore, they are not the same objects. Suppose now that we move from the abstract world of mathematics to the physical reality that surrounds us and let us assume that someone has drawn/constructed a material circle by putting minimal amounts of substance on what should be the circumference of the circle. Then, if we remove one amount of substance, is the result still a circle? Without question, this reminds us of the sorites paradox, but what is an amount of substance?

Nigel Wheatley discussed this problem in a short paper [72]. Chemists define an amount of substance n, which is called *mole*, as a quantity that contains exactly N_A elementary entities (e.g., molecules or atoms), where $N_A = 6.02214076 \times 10^{23} \, \text{mol}^{-1}$ is the Avogadro number.[8] This definition makes people to confuse the amount of substance with the number of entities. Thus, it is quite reasonable to ask whether $N_A - 1$ elementary entities make up a mole? It also reasonable to ask whether $N_A - 10$ elementary entities are a mole? The purely mathematical is answer is no, but outside the mathematical world of exactness, given two volumes of the same gas that contain $N_A - 1$ and $N_A - 10$ elementary entities, respectively,

---

[8] This number is the ratio of the number of elementary entitie in a substance to its amount of substance and is named after the Italian scientist Lorenzo Romano Amedeo Carlo Avogadro, Count of Quaregna and Cerreto.

at the same temperature and pressure, it is almost impossible to distinguish the two volumes. The problem arises because the amount of substance is a *continuous macroscopic quantity* whereas the number of entities is a *discrete microscopic quantity*. This simply means that we need new definitions that will address this problem. To return, to the original question about circle consisting of moles, it is obvious that it is again a sorites paradox.

A Précis of Basic Set Theory

2

A set is the result of combining into a whole definite, distinct objects of our intuition or thought — which are called the members of the set.

—*Georg Ferdinand Ludwig Philipp Cantor*
(German mathematician)

Paul Richard Halmos (Hungarian: Halmos Pál; 1916–2006) was a great mathematician of Hungarian origin who lived most of his life in the USA. In the preface of his "Naive Set Theory" [23] he starts by stating:

> Every mathematician agrees that every mathematician must know some set theory; the disagreement begins in trying to decide how much is some.

Since readers of this book definitely need to know something about sets and their properties, this chapter is my answer to the question of how much is some.

2.1 What Is a Set?

Even when one claims that she had never heard the term set before, it is more than sure that she has used the notion of a set many times in her everyday life. For instance, when we talk about a flock of seagulls, a herd of elephants, or a swarm of bees we actually talk about sets. Roughly, we could say that a set is a *collection*, a *group*, but not a *class* of things, animals, or, more generally, objects. In general, it is not easy or even possible to get more specific. Nevertheless, it should be obvious that all objects that make up a set should be of the same kind. But what is so special about classes that prohibits their use in the "definition" of sets? The answer is simple: This term has a precise technical meaning in mathematics, so it is best to avoid using this term when talking about sets.

© The Author(s), under exclusive license to Springer Nature Switzerland AG 2025
A. Syropoulos, *Fuzzy Mathematics*, Synthesis Lectures on Mathematics & Statistics,
https://doi.org/10.1007/978-3-031-73834-0_2

Many consider sets to be the most fundamental mathematical *structure*. In fact, there are approaches to mathematics that assume that everything in mathematics, including numbers, are sets. Here, we will not follow this slippery road and we will assume that numbers are objects that are different from sets. Of course, this choice has certain implications, but I do not think it makes sense to address them here.

Euclid (Εὐκλείδης) was a famous ancient Greek mathematicians. He is best known for his Elements [19] a collection of books that laid the foundations of geometry. In particular, he starts by stating a few axioms about points and lines and then, using these axioms, he proves other statements about lines, line segments, circles, etc. However, Euclid did not make any attempt to explain what is a point or a line. It seems that he thought that his readers would have an intuitive understanding of these concepts. In a way, the situation is quite similar when one tries to present sets, their properties and their operations—we do not have a precise definition of sets but we do have axioms that are used to define the properties and the operations between sets. Certainly, my purpose is not to present set theory in its full mathematical glory, but rather to present the basics in an understandable way.

The objects that make up a set are typically called its *elements* or its *members*. For example, when we talk about the set of pupils of a certain class, then Mike, Jim, Mary, Jenny, etc., are the elements of the set. Of course, we can have sets that have as elements other sets. Thus, the set of all classes of a school is actually a set whose elements are sets of pupils. Typically, we use capital letters to specify sets (i.e., A, B, etc.) and lowercase letters to specify elements of a set.

One of the most basic properties of a set is *membership* and it asks whether some element belongs or does not belong to a set. When the element x belongs to set A we write

$$x \in A.$$

But when the element y is not a member of the set A, then we write

$$y \notin A.$$

The use of this symbol was proposed by Giuseppe Peano and here is how it was introduced in [46, p. 156]:

> To indicate the singular proposition "x is an individual of the class s," we shall write[a]
>
> $$x \epsilon s$$
>
> and the sign ϵ may be read *is*, or *is a*, or *was*, or *will be*, according to grammatical rules, but its meaning is always that explained.
>
> ———————
>
> [a] The sign ϵ is the initial of ἐστί.

Clearly, the symbol ϵ is different from $\in$ but the later was, in a sense, the evolution of the former. The later was introduced in [10]. The ancient Greek word ἐστί means "is." Also, the reader who might be interested in the history of mathematical symbols is advised to have a look at Florian Cajori's book [13].

Sets can be *equal* or not equal. When two sets A and B are equal we write

$$A = B.$$

In case two sets are not equal, then we denote this by

$$A \neq B.$$

But when are two sets equal? Two sets are equal when the satisfy the following axiom:

Axiom 2.1.1 (*Axiom of extension*) Two sets are equal if and only if they have the same elements.

Traditionally, the *natural* numbers are those numbers that we use to count things, that is, the numbers $1, 2, 3, 4, \ldots$. Typically, we denote the natural numbers by $\mathbb{N}$. Some authors assume that zero is a natural number and so they use the symbol $\mathbb{N}_0$ to denote the set of natural number including zero. However, others use the symbol $\mathbb{N}$ to denote the set of natural numbers including zero. Since, this practice is followed by most modern authors, I will follow it for the rest of this text.

Integers are whole numbers (i.e., not fractional numbers) that can be positive, negative, or zero. The set of all integers is denoted by $\mathbb{Z}$ and the set of all positive integers is denoted by $\mathbb{Z}^+$. The set of integers *contains* the set of natural numbers since all natural numbers are integers but not all integers are natural numbers. Because of this property, we say that the set of natural numbers is a *subset* of the set of integer numbers and we denote this by

$$\mathbb{N} \subseteq \mathbb{Z} \quad \text{or} \quad \mathbb{Z} \supseteq \mathbb{N}.$$

The first relation is pronounced "$\mathbb{N}$ is a subset of $\mathbb{Z}$," whereas the second relation is pronounced "$\mathbb{Z}$ is a superset of $\mathbb{N}$." When A is not a subset of B, then we denote this by

$$A \nsubseteq B \quad \text{or} \quad B \nsupseteq A.$$

When we have two sets A and B such that $A \subseteq B$ and $A \neq B$, then we say that A is a *proper* subset of B and we denote this by

$$A \subset B \quad \text{or} \quad B \supset A.$$

For example, it is not difficult to see why $\mathbb{N} \subset \mathbb{Z}$. The set of all numbers that can be expressed as fractions whose numerator and denominator are integers, are called *rational* numbers and their set is denoted by $\mathbb{Q}$. The union of the set $\mathbb{Q}$ and the set of all *irrational* numbers, that is, all those numbers that cannot be expressed as a fraction of integers (e.g., the numbers $\sqrt{2}$ and $\sqrt{5}$ are such numbers), are called *real* numbers, and is denoted by the letter $\mathbb{R}$.

Suppose that A, B, and C are sets and it holds that $A \subseteq B$, $B \subseteq C$, then $A \subseteq C$. In mathematics this property is called *transitivity*. It is not difficult to see that for any three sets A, B, and C if $A = B$ and $B = C$, then $A = C$. Thus, set equality is also transitive relation. In addition, it holds that if $A = B$, then $B = A$. However, note that if $A \subset B$, then $B \not\subset A$.

2.2   Sets and Their Elements

Although I have said some basic things about sets, I have not said anything about the way we write a set and its elements. The most common way to write a set is by listing its elements between curly braces (i.e., the symbols { and }). For example, a set containing the numbers 1, 2, and 3 would be written as $\{1, 2, 3\}$. When we have a set of students, it is not difficult to write the set of all female students. However, things are not that easy when we talk about sets that may have an infinite number of elements. For example, try to write down the set of all odd natural numbers. If you have tried to write this set, I am sure you have already realized that you need an infinite amount of time to write all of its elements. Thus, we need a different method to write down such sets.

The set *builder* notation is used to write sets whose elements share some property. To specify a set using this notation, we write the general form of its elements, a vertical bar, and then the property that its elements share. For example, here is how can specify the set of all odd numbers

$$\{n \mid n = 2k + 1 \text{ and } k \in \mathbb{N}_0\}.$$

The set description that follows is the set containing the numbers 1, 2, and 3.

$$\{1, 2, 3\} = \{x \mid x \in \mathbb{N} \text{ and } 1 \leq x \leq 3\}.$$

The mathematical background of the notation just described is the following axiom

Axiom 2.2.1 (*Axiom of specification*) For every set A and every condition $S(x)$ there is a set B whose elements are exactly those elements x of A such that $S(x)$ holds true.

According to the axiom of specification it is possible to write down the set

$$B = \{x \mid x \in A \text{ and } x \notin x\}$$

and the question is whether $B \in A$? Suppose that $B \in A$. Then, this means that either $B \in B$ or $B \notin B$. Assume that $B \in B$. Then, $B \in A$ and $B \notin B$, which is a contradiction. Alternatively, if $B \notin B$, then $B \in A$ and not $B \notin B$, which means that $B \in B$ and this is again a contradiction. The result of this mental exercise, which mathematicians call *proof*, is that *nothing contains everything*. This is a very important conclusion, because in what follows I will talk about a *universe* (i.e., a set of everything) but we have just shown that such a set does not exist. Thus, when we talk about a universe we need to be careful enough to mean a set whose elements are all those general elements that are of interest to us for a particular case.

A third method to describe subsets of some set, which is usually called a universe, employees the *characteristic* or *indicator* function. Roughly, a function is a mechanism to relate the elements of a set A to the elements of a set B. The set A is called the *domain* of the function and the set B its *codomain* or *range*. In mathematics, lowercase letters are usually used for functions (e.g., f, g, h, etc.). We can view a function as a set of *ordered pairs*, that is, a set that has elements of the form (a, b), which is something quite different from the set $\{a, b\}$ [26] (in Sect. 2.6 I will say more about ordered pairs). The elements of $\{a, b\}$ are assumed to be different and the order they occur is irrelevant. While in (a, b) a can be equal to b and $(a, b) \neq (b, a)$. A function f whose domain is the set X and whose codomain is the set Y is written as $f : X \to Y$. Suppose that

$$A = \{\text{James, Robert, John, Michael}\}$$

and

$$B = \{\text{Emma, Charlotte, Amelia}\}.$$

Then, the function "likes" that models which boy likes which girl in a given classroom can be defined as follows:

$$\text{likes}(\text{James}) = \text{Emma}$$
$$\text{likes}(\text{Robert}) = \text{Emma}$$
$$\text{likes}(\text{John}) = \text{Charlotte}$$
$$\text{likes}(\text{Michael}) = \text{Charlotte}$$

A visual representation of this function definition is given in Fig. 2.1. The fact that Robert likes Emma is expressed by drawing an arrow from the element Robert to the element Emma. Note that every boy should like at most one girl. If some boy likes two or more girls, then the resulting structure is not a function. Let us now describe function likes as a set:

$$\text{likes} = \{(\text{James, Emma}), (\text{Robert, Emma}), (\text{John, Charlotte}), (\text{Michael, Charlotte})\}.$$

If every boy likes either no girl from the list or only one and no other boy likes this same girl, then the resulting function is *injective*. If for every girl there is one or even more boys that like her, then resulting function is *surjective*. If every boy likes one and only one girl and no other boy likes the same girl, the then resulting function is *bijective*. In different words, any function that is both injective and surjective, is a bijective function.

If we have a bijective function f, then we find its *inverse*, that is another function g that takes as argument the value $f(a)$ and computes a. For example, consider the function $f(x) =$

Fig. 2.1 A visual
representation of function
"likes"

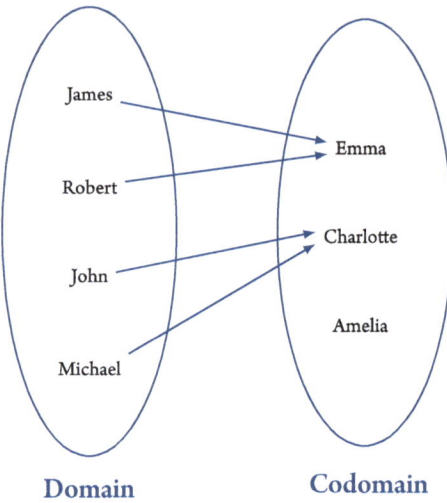

5x − 1 and the value $f(3) = 5 \times 3 - 1 = 15 - 1 = 14$. Then, function $g(x) = (x + 1)/5$
is the inverse of f: $g(14) = (14 + 1)/5 = 15/5 = 3$. The reader can experiment with other
values and verify that g is indeed the inverse of f. In mathematics, the symbol f^{-1} denotes
the inverse of a function f. The next reasonable question is: Is there an easy way to find the
inverse of a function? Fortunately, the answer is "yes." In particular, when we want to find
the inverse of some function, we need to solve an algebraic equation. For example, assume
we wan to find the inverse of $f(x) = 5x - 1$. Then, we put "y" for "$f(x)$" and solve for
"x:"

- We start with $f(x) = 5x - 1$.
- We put "y" for "$f(x)$:" $y = 5x - 1$.
- Add 1 to both sides: $y + 1 = 5x$.
- Divide both sides by 5: $\frac{y+1}{5} = x$.
- Swap sides: $x = \frac{y+1}{5}$.
- Put "$f^{-1}(y)$" for "x" to get the solution: $f^{-1}(y) = \frac{y+1}{5}$.

Let us now define the characteristic function of a set:

Definition 2.2.1 Assume that X is a set called a universe.[1] Also, suppose that $A \subseteq X$. Then,
its *characteristic* function is is a function $\chi_A : X \to \{0, 1\}$ of A that is defined as follows:

$$\chi_A : X \to \{0, 1\} \text{ such that: for every } x \in X : \chi_A(x) = 1 \text{ if and only if } x \in A.$$

---

[1] A non-rigorous description of this set will be given in Sect. 2.5.

A more compact way to say the same thing is the following expression:

$$\chi_A(a) = \begin{cases} 1, \text{ if } a \in A, \\ 0, \text{ if } a \notin A. \end{cases}$$

Example 2.1 The characteristic function of the set that corresponds to function "likes" is defined as follows:

$$\chi_{\text{likes}}(\text{James, Emma}) = 1$$
$$\chi_{\text{likes}}(\text{Robert, Emma}) = 1$$
$$\chi_{\text{likes}}(\text{John}), \text{Charlotte}) = 1$$
$$\chi_{\text{likes}}(\text{Michael, Charlotte}) = 1$$

As a second example, assume that E is the set of positive even natural numbers. Then, its characteristic function can be defined as follows:

$$\chi_E(n) = \frac{1 + (-1)^n}{2}.$$

Similarly, the characteristic function of the set of odd natural numbers can be defined as follows:

$$\chi_O(n) = \frac{1 + (-1)^{n+1}}{2}.$$

2.3   The Empty Set

Suppose that A is a set. Then, according to the axiom of specification the following set

$$\{x \mid x \in A \text{ and } x \neq x\}$$

exists and it is unique. In particular, this set is called the *empty* and is denoted by $\varnothing$ or by {}. In addition, the set $\varnothing$ is a subset of every set, that is,

$$\varnothing \subset A, \text{ for any set } A.$$

In order to show that this statement is true, one has to prove that it cannot be false. In different words, can we show that $\varnothing \subset A$ is false? This statement can be false only if the set $\varnothing$ contains an element that is not an element of A. We know that the set $\varnothing$ has no elements. Therefore, the statement $\varnothing \subset A$ cannot be false and so it must be true for every A.

It is rather interesting but with the ideas presented so far, we have managed to prove the existence of the empty set only. Clearly, there are more sets, even an infinite number of sets, but how can we ascertain their existence? The following axiom is a good starting point for the introduction of more sets.

Axiom 2.3.1 (*Axiom of pairing*) For any two sets there exists a set that they both belong to.

In simple words, this axiom says that if A and B are sets, then there exists a set C such that $A \in C$ and $B \in C$. This set can be expressed using the set builder notation as follows:

$$\left\{x \mid x \in C; \text{ such that } x = A \text{ or } x = B\right\}.$$

Now, the axiom implies that this set, which is usually written as $\{A, B\}$, is unique. If A is a set, then we can also form the set $\{A, A\}$, which is denoted by

$$\{A\},$$

which is the *singleton* and contains only one element. From this we can easily conclude that $\varnothing$ and $\{\varnothing\}$ are two different things. The first is the empty set and the second a singleton that contains the empty set.

2.4   Unions and Intersections of Sets

Math teachers use a graphical method to draw sets and the result of operations between sets. This method was devised by the English logician and philosopher John Venn (1834–1923). Sets are represented by circles that contain bullets whose labels are the elements of the set. For example, the set $A = \{1, 2, 3\}$ can be represented by the following diagram.

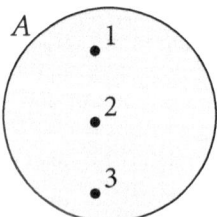

Suppose that A and B are two sets. Then, their *union* is a new set that contains all the elements which are in A, in B, or in both A and B. In fact, the existence of the union of two sets is based on the validity of the following axiom.

Axiom 2.4.1 (*Axiom of unions*) For every collection of sets there exists a set that contains all the elements that belong to at least one set of the given collection.

Assume that the letter $\mathscr{C}$ stands for the collection of sets and U is the set that contains all the elements of all set X that belong to $\mathscr{C}$. Then, U is defined as follows.

$$U = \{x \mid x \in X \text{ for some } X \in \mathscr{C}\}.$$

The set U is called the union of all sets X that make up the collection $\mathscr{C}$ and we use the expression that follows to denote this property.

$$U = \bigcup_{X \in \mathscr{C}} X$$

This *definition* leads naturally to the following facts:

$$\bigcup_{X \in \varnothing} X = \varnothing \text{ and } \bigcup_{X \in \{A\}} X = A.$$

Now, consider the set

$$\bigcup_{X \in \{A, B\}} X = \{A, B\}.$$

This is the union of two sets and, as was stated above, it contains all the elements which are in A, in B, or in both A and B. Thus,

$$A \cup B = \{x \mid x \in A \text{ or } x \in B\}.$$

Here "or" means that either $x \in A$ or $x \in B$ is true. The following statements can be proved easily.

$$A \cup \varnothing = A$$
$$A \cup B = B \cup A$$
$$A \cup (B \cup C) = (A \cup B) \cup C$$
$$A \cup A = A$$
$$A \subset B \text{ iff } A \cup B = B.$$

The word *iff* means *if and only if*

Suppose that $A = \{1, 2, 3, 4, 5\}$ and $B = \{4, 5, 6, 7\}$ are two sets. Then, their union is depicted by the following Venn diagram.

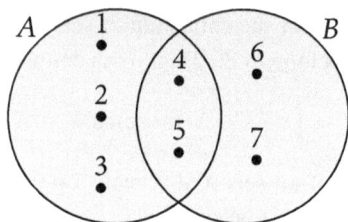

Note that the overlapping area is reserved for the elements that are common to both sets.

If we have two sets A and B, their *intersection* is the set $A \cap B$ that contains those elements that are *common* to both sets.

$$A \cap B = \{x \mid x \in A \text{ and } x \in B\}.$$

Here "and" means that both $x \in A$ and $x \in B$ are true. The following statements can be proved easily.

$$A \cap \emptyset = \emptyset$$
$$A \cap B = B \cap A$$
$$A \cap (B \cap C) = (A \cap B) \cap C$$
$$A \cap A = A$$
$$A \subset B \text{ iff } A \cap B = A.$$

Consider again the sets $A = \{1, 2, 3, 4, 5\}$ and $B = \{4, 5, 6, 7\}$. Then, their intersection is depicted by the following Venn diagram.

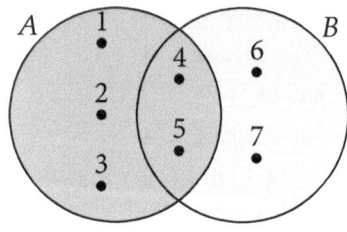

Here the overlapping area is the intersection of two two sets.

The following equalities that involve both unions and intersection are quite useful.

$$A \cap (B \cup C) = (A \cap B) \cup (A \cap C),$$
$$A \cap (B \cap C) = (A \cup B) \cap (A \cup C).$$

The formation of the intersection of two sets A and B or, alternatively, the formation of the intersection of a pair $\{A, B\}$ of sets can be formally defined in a way similar to the definition of the union of two sets and is left as an exercise for the reader.

2.5   Set Difference and Symmetric Difference

The *difference* of two sets A and B is the set $A \setminus B$ defined by

$$A \setminus B = \{x \mid x \in A \text{ and } x \notin B\}.$$

Consider again the sets $A = \{1, 2, 3, 4, 5\}$ and $B = \{4, 5, 6, 7\}$. Then, their difference is depicted by the following Venn diagram.

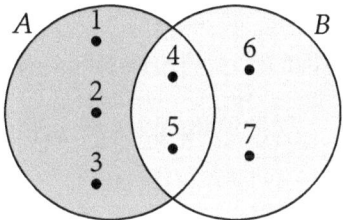

Here the darker area is the result of the operation.

Suppose now that there is a set U, which will be called the *universe* (or universal), which contains all elements of a (big) group of sets. Then, the *complement* of a set A, which is a member of this group of sets, is denoted by A^C and it is the set $U \setminus A$. The diagram that follows shows a set and its complement.

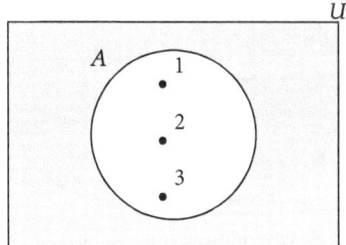

The gray area is the complement of set A. "Ordinary" set theory does not allow for the existence of a universe. However, there are some set theories that allow for the existence of this set by admitting the following axiom [30]:

Axiom 2.5.1 (*Axiom of the universe*) The set

$$U = \{x \mid x = x\}$$

exists.

The following are the basic properties of complementation:

$$(A^{\complement})^{\complement} = A$$
$$\varnothing^{\complement} = U$$
$$U^{\complement} = \varnothing \tag{2.1}$$
$$A \cap A^{\complement} = \varnothing$$
$$A \cup A^{\complement} = U$$

In addition, $A \subset B$ if and only if $B^{\complement} \subset A^{\complement}$. The following very important properties are called *De Morgan laws*[2]:

$$(A \cup B)^{\complement} = A^{\complement} \cap B^{\complement} \text{ and}$$
$$(A \cap B)^{\complement} = A^{\complement} \cup B^{\complement}. \tag{2.2}$$

Assume that A and B are sets. Then, their *symmetric difference* is the set $A \triangle B$ defined by

$$A \triangle B = (A \setminus B) \cup (B \setminus A).$$

If A and B are the sets we have used so far in our examples, then their symmetric difference is depicted by the following Venn diagram.

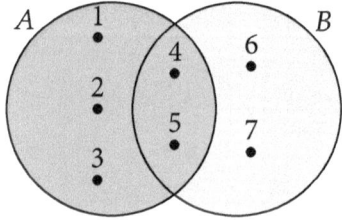

Here the darker area is the result of the operation.

For every set U there is a special set whose existence is guaranteed by the following axiom:

Axiom 2.5.2 (*Axiom of powers*) For each set there exists a collection of sets that contains among its elements all the subsets of the given set, including the empty set.

For a set A, this set is know as the *power* set and is denoted by $\mathscr{P}(A)$. Sometimes, the power set of a set A is denoted by 2^A. In general, the notation B^A denotes a set of functions:

$$B^A = \{f : A \to B\}$$

---

[2] A law is a mathematical statement that always holds true.

Thus, the notion 2^A denotes the set of all functions from A to a given set of two elements (e.g., $\{0, 1\}$). The power set of A is equivalent to set of all the functions from A to the given two-element set. Practically, this means that if $B \subset A$, then B corresponds to the following function

$$f(x) = \begin{cases} 1, & \text{if } x \in B \\ 0, & \text{if } x \notin B \end{cases}$$

Here we assumed that $2 = \{0, 1\}$ and this comes from John Von Neumann's definition of the natural numbers:

$$0 = \{\} = \varnothing,$$
$$1 = \{0\} = \{\varnothing\},$$
$$2 = \{0, 1\} = \{\varnothing, \{\varnothing\}\},$$
$$3 = \{0, 1, 2\} = \{\varnothing, \{\varnothing\}\{\varnothing, \{\varnothing\}\}\}$$

Von Neumann's numerals cannot and should not be identified with the natural numbers. Although, individual von Neumann numerals exist in general, we do not really know if the set of all von Neumann numerals exist.

Example 2.2 Assume that

$$A = \{1, 2, 3\}.$$

Then,

$$\mathscr{P}(A) = \{\varnothing, \{1\}, \{2\}, \{3\}, \{1, 2\}, \{2, 3\}, \{1, 3\}, \{1, 2, 3\}\}.$$

∎

2.6   Cartesian Product of Sets

In sets order does not matter. This means that the order in which we specify the elements of some set is irrelevant. Thus, the sets

$$\{1, 2, 3, 4\} \quad \text{and} \quad \{4, 2, 1, 3\}$$

are identical. Suppose now that for some weird reason we want to define a set where the order of the elements does matter. Here "order" can mean anything that puts the elements of a set in some order. For example, consider a set with three elements a, b, and c that are ordered as follows:

$$cba.$$

One way to create a set that *encodes* this order in its structure is to create a set whose elements
are the sets

$$\{c\}, \{c, b\}, \text{ and } \{c, b, a\}.$$

How can we extract the order from this set? Here is a recipe: The singleton set contains the
element that is first in the order. By removing the element of the singleton set from the set
that has two elements, we get the second element in the order. By following this procedure,
we can extract the elements in the specific order. Using this recipe, we can define the ordered
pair of a and b, with first coordinate a and second coordinate b, to be the set (a, b) defined by

$$(a, b) = \{\{a\}, \{a, b\}\}.$$

This definition is correct only if the basic property of ordered pairs is valid. That is, if (a, b)
and (c, d) are ordered pairs and if $(a, b) = (c, d)$, then $a = c$ and $b = d$. Assume that $a = b$.
Then, the (a, b) is actually the following set

$$\{\{a\}, \{a, a\}\} = \{\{a\}, \{a\}\} = \{\{a\}\}.$$

On the other hand, if (a, b) is a singleton, then $\{a\} = \{a, b\}$, which means that $b \in \{a\}$ and
so $a = b$. Assume that $(a, b) = (c, d)$. Then, if $a = b$ we deduce that (a, b) and (c, d) are
singletons and it follows that $a, b, c,$ and d are all equal. If $a \neq b$, then we know that (a, b)
contains a singleton, viz. $\{a\}$, and (c, d) contains also a singleton, viz. $\{c\}$. Since the singletons must be equal it follows that $a = c$. In addition, both (a, b) and (c, d) contain one
unordered pair, viz. $\{a, b\}$ and $\{c, d\}$, respectively. This implies that $\{a, b\} = \{c, d\}$. This
means that $b \in \{c, d\}$ which implies that $b = d$, since it cannot be equal to c.

The next question we need to answer is: if A and B are sets, is there a set that contains
all the ordered pairs (a, b) such that $a \in A$ and $b \in B$? The answer is yes and here is
a proof of it. Suppose that $a \in A$ and $b \in B$. Then, $\{a\} \subset A$ and $\{b\} \subset B$, which means
that $\{a, b\} \subset A \cup B$. In addition, $\{a\} \subset A \cup B$, which means that $\{a\} \in \mathscr{P}(A \cup B)$ and
$\{a, b\} \in \mathscr{P}(A \cup B)$. This simply means that

$$\{\{a\}, \{a, b\}\} \subset \mathscr{P}(A \cup B) \text{ and so } \{\{a\}, \{a, b\}\} \in \mathscr{P}\big(\mathscr{P}(A \cup B)\big).$$

What is left is to apply the axioms of specification and extension to create the unique set
$A \times B$ that contains all the ordered pairs (a, b) such that $a \in A$ and $b \in B$. The set $A \times B$
is called the *Cartesian product* of A and B. This set is characterized by the following fact:

$$A \times B = \{(a, b) \mid a \in A \text{ and } b \in B\}.$$

Example 2.3 Let $A = \{1, 2, 3, 4\}$ and $B = \{6, 7, 8\}$ be two sets. Then, their Cartesian product is the set

$$A \times B = \{(1, 6), (1, 7), (1, 8), (2, 6), (2, 7), (2, 8), (3, 6), (3, 7), (3, 8), (4, 6), (4, 7), (4, 8)\}.$$

■

The next few facts can be proved easily. If $A = \varnothing$ and $B \neq \varnothing$, then $A \times B = \varnothing$. Also,

$$(A \cup B) \times C = (A \times C) \cup (B \times C),$$
$$(A \cap B) \times (C \cap D) = (A \times C) \cap (B \times D),$$
$$(A \setminus B) \times C = (A \times C) \setminus (B \times C).$$

Fuzzy Subsets

3

Gothic cathedrals and Doric temples are mathematics in stone

—Oswald Spengler
(German philosopher)

In the previous chapter I presented the core ideas of set theory. The exposition was not very formal, nevertheless, it was rigorous enough for our purposes. It is easy to see that set theory assumes that there is no vagueness. However, in the first chapter, I introduced vagueness and explained why vagueness is not just linguistic phenomenon. The real question is if we can define some sort of alternative mathematics where vagueness plays a central role. In what follows, I will explain what what is alternative mathematics and how we can introduce vague mathematics (and not mathematics of vagueness).

3.1 Alternative Mathematics

In the previous chapter, I presented the set of natural numbers (see Sect. 2.1) and I explained that some authors do not consider zero a natural number mainly because we use these numbers to count and zero is not used in counting. Surprisingly, the ancient Greeks did not consider zero a number and this position affected the development of Greek mathematics. But why did they not consider zero to be a number?

Ancient Greeks defined numbers to be entities that can be drawn [57, p. 49]. Thus, five is a number because we can draw a line whose length is 5 spithames [1 spithame is roughly 231.2 mm (9.10 in)]. But there is no line whose length is zero! So this makes really sense, although one may consider a point to be a line whose length is zero. However, the ancient Greeks had other more strange ideas. For example, the ancient Greek mathematicians

Archytas and Eudoxus did not consider the fourth power, the fifth power, the sixth power, etc., as real powers. This means that 7^4 was, according to them, something not real! The reason was simple: They associated the second powers with areas and the third powers with volumes and this is why they called them surface-numbers and volume-numbers, respectively. Thus, a fourth power does not correspond to something real and so it cannot exist. But the ancient Greeks did not stopped there. They believed that one is not a number and that it is neither odd nor even but even-odd [9, p. 110]. In addition, they believed that two is not an even number. Nowadays, most people would consider these ideas as nonsense. But why did the Greeks believe these things?

To them, one is the starting point or generator of numbers and so it cannot be a number. The following excerpt from Aristotle's *Metaphysics* explains why one is not a number.

> [1088a] [1] of qualities as a quality, and of quantities as a quantity. (The measure is indivisible, in the former case in kind, and in the latter to our senses.) This shows that unity is not any independent substance. And this is reasonable; because unity denotes a measure of some plurality, and number denotes a measured plurality and a plurality of measures. (Hence too it stands to reason that unity is not a number; for the measure is not measures, but the measure and unity are starting-points.) The measure must always be something which applies to all alike; e.g., if the things are horses, the measure is a horse; if they are men, the measure is a man; and if they are man, horse and god, the measure will presumably be an animate being, and the number of them animate beings. If the things are "man," "white" and "walking," there will scarcely be a number of them, because they all belong to a subject which is one and the same in number; however, their number will be a number of genera, or some other such appellation.[1]

Not everyone accepted the idea that one is not a number. In particular, Chrysippus, who lived in the 3rd century B.C., called one as "multitude one" (πλῆθος ἕν). However, the Arab philosopher Iamblichus considered this statement a contradiction in terms [27, p. 69]. Because one is the generator of both even and odd numbers, this number cannot be odd and since it is not even it should be called even-odd. According to Theon of Smyrna, Aristotle considered the fact that when one is added to an even number the result is an odd number, but when added to an odd number the result is an even number. Thus, one is both an even and an odd number [27, p. 71]. To the Greeks, the number two (the dyad) was not an even number because just like the number one was the generator of numbers, two was the generator of even numbers.

The Greek word for number is *arithmos* (ἀριθμός or just αριθμός in modern Greek) and this is a word that is common to both ancient and modern Greek. However, the meaning of the word has changed. As Jacob Klein correctly notes in ancient Greek the word meant "counting, or more exactly, the *counting-off*, of some number of things" [35, p. 46]. In particular, for the ancient Greeks what was important was the objects they counted. Thus, when they counted apples, the result was a number of apples, but when the counted apples and figs, the result was a number of fruits, and when they counted apples, figs, and plates, then the result was

---

[1] Aristotle, *Metaphysics*, book 14, section 1088a, Perseus Digital Library, Tufts University.

a number of objects. And this is why Klein translated the word arithmos into German as *anzahl*, which means "a number of [things]," and not as *zahl*, which means just "number." Today, when modern Greeks use the word arithmos, they may mean a sequence of digits that corresponds to their social security number, which does not correspond to a quantity, or to their telephone number, etc. Certainly, numbers are also used to count things, but as is obvious the word has now a broader meaning. A direct consequence of this difference, is that ancient mathematics does not mean what we think it means. In fact, as Klein remarked, we actually reinterpreted the writings of the ancient Greeks in order to understand them. If one thinks a bit deeper, she will understand that if ancient Greek mathematics was the orthodox mathematics, then modern mathematics is in fact *alternative* mathematics, that is, mathematics where the core ideas have different meanings from the orthodox version. Thus, by changing the meaning of core ideas of mathematics, we can construct alternative mathematics. For example, this happened when Luitzen Egbertus Jan Brouwer who proposed that mathematical objects are mental constructs and not objects that "live" in a world of ideas, as Plato believed. This objection created *mathematical intuitionism* [29] a form of alternative mathematics. But was this objection enough to justify this?

In ordinary mathematics we are entitled to say that "this pair of trousers is yellow or this pair of trousers is not yellow" and statements like these are always true. This happens because something has or does not have a given property. However, in intuitionistic mathematics we are not allowed to make such statements. Assume that we have established the truth of the statement "this pair of trousers is yellow." Practically, this means that we have constructed a mental object (a proof) that this statement is true. Since we have done so, we cannot construct a mental object (a proof) of the statement "this pair of trousers is not yellow" because this is absurd! In simple words, this means that the statement above cannot be possible be true in intuitionistic mathematics. In classical mathematical logic, the validity of statements like the one above is called *the law of the excluded middle*. The morale of this is that any "violation" of the principles of classical mathematics can be used to create alternative mathematics.

3.2   What Is a Fuzzy Subset?

In ordinary mathematics everything is either *true* or *false*. Thus, the statement $x \in X$ is true if x is an element of the set X and false if x is not an element of X. Thus, when we have a set of students we can ask questions like "Is Maria a member of the basketball team?" or "Is Steve a member of the football team?". As expected, we can answer these questions with a "Yes" or a "No." Let us now ask if Maria is a tall girl or if Steve is a good player. As was explained in Chap. 1, we cannot say for sure whether Maria is a tall girl or if Steve is a good player. However, we can say something like "Maria is a very tall girl" or that "Maria is a quite a tall girl," etc. Similarly, we can say that "Steve is not a very good player" or that "Steve is a good player," etc. Since expressions like *very good*, *quite tall*, etc., can be understood in different ways by people, Zadeh and, independently, Dieter Klaua (e.g., see [34] but also [22] for a thorough presentation of Klaua's work) thought that it is better to use numbers that would numerically represent expressions like *very good*, etc. However, there is a subtle difference between the two approaches—Zadeh invented a mathematical tool to describe vagueness and uncertainty while Klaua extended classical mathematics as a result of a mathematical exercise or curiosity. Zadeh and Klaua proposed the use of the same numbers to arithmetically denote the various types of adjectives (i.e., comparative and superlative adjectives and their variations) that we use to express various properties. These numbers are drawn from the set [0, 1]. This set is known as the *unit interval*. More generally, an *interval* is a subset of the set of real numbers that contains all real numbers that are between two specific numbers. The most common forms of intervals are the following ones:

$$(a, b) = \{x \mid x \in \mathbb{R} \text{ and } a < x < b\}$$
$$[a, b] = \{x \mid x \in \mathbb{R} \text{ and } a \leq x \leq b\}$$
$$(a, b] = \{x \mid x \in \mathbb{R} \text{ and } a < x \leq b\}$$
$$[a, b) = \{x \mid x \in \mathbb{R} \text{ and } a \leq x < b\}$$

Thus, [0, 1] is the set of all real numbers that are less than or equal to one and greater than or equal to zero. In addition, Klaua proposed an alternative method to describe these words. In particular, he proposed a method where we choose a natural number $m \geq 2$ and then we construct a set that contains all the numbers $\frac{k}{m-1}$, where $0 \leq k \leq m - 1$. Obviously, the bigger the number m, the finer the *granularity*. Nowadays, most people prefer the first choice, so we will assume that the numbers belong to the unit interval. In this case, zero will mean that the given person does not have this property while one will mean that a person has this property to the fullest. Thus, we can say that "Maria is tall to degree equal to 0.75" or that "Steve is a good player to degree equal to 0.5," etc. Note that "Maria is tall to degree equal to 0.75" is completely different from "Maria is 75% tall." The later statement can be either true or false. But the former statement is a *compound* statement that consists of the statement "Maria is tall" and its denotation that happens to be the number 0.75. This is why the two statements are quite different.

Let us consider a class of students. Then, our task will be to determine to what degree each student of this class is tall. After a careful evaluation, we will end up with statements like "Olivia is tall with degree equal to 0.50," "Amelia is tall with degree equal to 0.45," etc. Let us make our results more formal (i.e., more mathematical). The following is a mathematical way to compactly write statements like these.

$$\text{tall(Olivia)} = 0.50$$
$$\text{tall(Amelia)} = 0.45$$
$$\text{tall(Emma)} = 0.75$$
$$\text{tall(Michael)} = 0.80$$
$$\vdots$$

Without a second thought, one can say that these mathematical expressions define a function whose domain is the set of students and its codomain is the unit interval. In fact, one can think of this function as the analog of the characteristic function. Since the characteristic function is used to define sets, this function could be used to define a new kind of set—a set whose members belong to it to some degree. Interestingly, as was noted in the previous paragraph, when an element has a *membership degree* equal to zero, the element belongs to the structure with degree equal to zero. These new structures are called fuzzy subsets.

Definition 3.2.1 Let X be a *universe* (i.e., an arbitrary set). A *fuzzy subset A* of X, is realized by a function $A : X \rightarrow [0, 1]$, which is called its *membership function*. For every $x \in X$, the value $A(x)$ is the *degree to which element x belongs to the fuzzy subset A*.

At this point it is necessary to make a few remarks. First, some authors still use the old notation $\mu_A(x)$ to write the membership degree of the element x to the fuzzy subset A. Second, most people talk about fuzzy sets when in fact the correct term is fuzzy subset. Third, the universe of the definition above should not be confused with the set of everything that we called universe in the previous chapter. From the definition of intervals, the set $[0, 1]$ contains numbers like $e/8$,[2] $\sqrt{2}/4$, and so on, that can be used as membership degrees. In the first chapter, it was proposed to use approximations of these numbers but a more mathematically correct definition would be to say that membership degrees should be elements of the set $\mathbb{Q} \cap [0, 1]$, that is, all the rational numbers that belong to the unit interval.

Given a universe x, a *normal* fuzzy subset A of X is one for which there is at least one element $y \in X$ such that $A(y) = 1$. If this condition is not met, then A is called subnormal. In practical terms, a normal fuzzy subset is one where at least one element has the specific property to the fullest. Thus, if

$$\text{tall}(\text{Roger}) = 1,$$

then this means that Roger is really tall and on one can dispute this.

As was explained above, a fuzzy subset A of some ordinary set X is a function, but when X is relatively small set, then we can write A as follows:

$$A = \{a_1/x_1, a_2/x_2, \cdots, a_n/x_n\},$$

where $x_i \in X$, a_i is the degree to which x_i belongs to A, and a_i/x_i means that x_i belongs to A with a degree that is equal to a_i. Seeing that notation, one may wrongly conclude that a fuzzy subset is actually an ordinary set. Of course, this is wrong because we *overload* the notation, which simply means that the notation gets an additional meaning. This a very common practice in computer programming and it is used also in mathematics. Some authors prefer the following notation

---

[2] The symbol e denotes Napier's constant (named after John Napier of Merchiston), which is also known as Euler's number (named after the great Swiss mathematician Leonhard Euler). It is roughly equal to 2.71828 and it is an irrational number.

$$A = a_1/x_1 + a_2/x_2 + \cdots + a_n/x_n,$$

but I believe the previous one is closer to the idea of a set. Again, "+" does not mean addition but concatenation (i.e., putting things one after the other). Alternatively, one can write down a fuzzy subset as follows:

$$A = \sum_{i=1}^{n} A(x_i)/x_i,$$

here x ranges over a set of discrete values. Let us now see how we can write the fuzzy subset of tall people using the previous notation.

$$\text{tall} = \left\{ \frac{0.50}{\text{Olivia}}, \frac{0.45}{\text{Amelia}}, \frac{0.75}{\text{Emma}}, \frac{0.80}{\text{Michael}}, \ldots, \frac{1.00}{\text{Roger}} \right\}.$$

So far we have discussed fuzzy subsets of finite sets, but is it possible to have fuzzy subsets of infinite sets? From a purely mathematical point of view, there is no problem to define such a set. However, in this case we need a formula to automatically assign a membership degree to each element of the infinite sets. In cases like these, a fuzzy set can be written down as

$$A = \int_X a_i/x_i.$$

Again, the symbol $\int$ does not denote integration. Also, x ranges over a continuum.

3.3 α-Cuts of Fuzzy Subsets

Suppose we have a set of students and the fuzzy subset of good students. As is known, at the end of the year, each student gets a grade for her overall level of achievement for all courses she attended. Suppose that we use the following scale:

Description	Grade
Fail	1–9.5
Marginal performance	10–12
Satisfactory	12.5–15
Very well	15.5–18
Extremely well	18.5–20

Then, we can construct a fuzzy subset of good students by *normalizing* the scale above so that each grade is a number that belongs to the unit interval. If we want to find the students whose grade is "marginal performance," then we have to find an α-cut (pronounced alpha-cut) of the fuzzy subset of good students. These sets are also known in the literature as α-level and they are *crisp* sets (i.e., ordinary sets). In our case, they contain all students whose grade is greater than or equal to some threshold grade. For example, consider the following fuzzy subset of a set of students.

$$A = \left\{ \frac{0.70}{\text{Steve}}, \frac{0.80}{\text{Marcy}}, \frac{0.95}{\text{Al}}, \frac{0.40}{\text{Peggy}}, \frac{0.85}{\text{Bud}}, \frac{0.48}{\text{Kelly}} \right\}.$$

Suppose α is 0.50, then the α-cut of A is the set {Steve, Marcy, Al, Bud} because all of these students have a grade (i.e., a membership value) greater than 0.50. Formally, we define the α-cut as follows:

Definition 3.3.1 Suppose that $A : X \to [0, 1]$ is a fuzzy subset of X. Then, for any $\alpha \in [0, 1]$, the α-cut $^{\alpha}A$ and the strong α-cut $^{\alpha+}A$ are the crisp sets

$$^{\alpha}A = \left\{ x \mid x \in X \text{ and } A(x) \geq \alpha \right\}$$

and

$$^{\alpha+}A = \left\{ x \mid x \in X \text{ and } A(x) > \alpha \right\},$$

respectively.

A special kind of α-cut is the *support* of a fuzzy subset. In particular, the support of a fuzzy subset $A : X \to [0, 1]$, denoted by supp(A), is the set of all elements of X that have nonzero membership degrees, that is, the support of A is its strong α-cut ^{0+}A. Figure 3.1 shows a fuzzy subset, the α-cut when $\alpha = 0.7$, it support and its *core* (i.e., the set of all x that have membership degree equal to 1).

An interesting property of α-cuts is that for any fuzzy subset $A : X \to [0, 1]$ if $a_1 < a_2$, then

$$^{a_1}A \supseteq ^{a_2}A \quad \text{and} \quad ^{a_1+}A \supseteq ^{a_2+}A.$$

Fig. 3.1 The fuzzy subset B (solid line), an α-cut (dashed line), its support, and its core

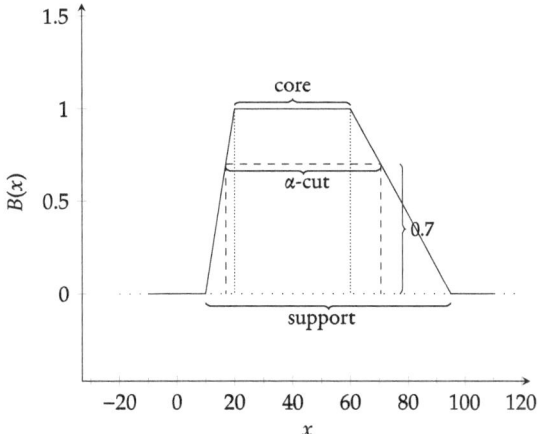

In addition, if A and B are two fuzzy subsets of X, then for all $a, b \in [0, 1]$ the following properties hold true:

(1) ${}^{a+}A \subseteq {}^{a}A$;
(2) ${}^{a}(A \cap B) = {}^{a}A \cap {}^{a}B$ and ${}^{a}(A \cup B) = {}^{a}A \cup {}^{a}B$;
(3) ${}^{a+}(A \cap B) = {}^{a+}A \cap {}^{a+}B$ and ${}^{a+}(A \cup B) = {}^{a+}A \cup {}^{a+}B$; and
(4) ${}^{a}(A^{C}) = ({}^{(1-a)+}A)^{C}$.

A very important result (see [36] for its proof) is the following:

Theorem 3.3.1 *For every fuzzy set $A : X \to [0, 1]$,*

$$A = \bigcup_{\alpha \in [0,1]} {}^{\alpha}A,$$

that is, the fuzzy set A is equal to the union of all its α-cuts.

3.4   How Do We Assign Membership Degrees?

So far I have described what is a fuzzy subset and how one can write it down. However, I have not discussed how we do assign membership degrees to elements of a fuzzy subset (in fact, I briefly discussed this matter in Sect. 1.3, but there I talked about vagueness, in general). From the discussion so far, one may get the impression that we should use our intuition and our knowledge of things to *subjectively* assign membership degrees to elements of a fuzzy subset. Although, this is usually true, there are techniques and tools that can be employed to assign reasonable membership degrees to elements of fuzzy subsets. Since the "problem of vagueness exists in natural and social sciences, but it appears more seriously in the social sciences" [44], I will present methods and ideas that are used primary in social sciences [56, 70].

A membership degree is only one number that is associated with an object x. This means that it ignores other characteristics (or dimensions, if you prefer this term) of x. Thus, we need to define more fuzzy subsets to handle these additional characteristics or we can can define a far more general form of fuzzy subsets that can describe all of them as once. I will discuss this solution later but for now I will continue by assuming that a membership degree is just a number that belongs to the set [0, 1]. In [56] it is claimed that "membership is latent, that is, not directly observable." I do not agree with this statement since when we define a fuzzy set we *see* the objects and the membership degrees reflect what we see. But what we see depends on the *context*. For example, if we are to assign membership degrees and think of NBA players, then of course a boy whose height is 1.80 cm (5 feet and 11 inches) tall is definitely a short person. To make things even worse, I asked Google to tell me whether a person whose height is 1.80 cm is a tall person and here is what I got:

With 180 cm you're at average height or slightly below in around 20 countries, and in all other countries you are above. The Netherlands is the country with the tallest people, averaging 183 cm for men. So yeah, 180 cm is tall in most countries, average in a few, and definitely not short anywhere.

Obviously, this implies that when we define a fuzzy set, we have to specify the context as clearly as possible.

In general, people prefer to use verbal phrases instead of membership degrees. Thus, it is easier to say that "John is not tall at all" instead of saying that John's membership degree to the set of tall people is 0.0 or that "John is tall with degree equal to 0.0." In addition, it is easier to say that "John is absolutely tall" to express the fact that "John is tall with degree equal to 1.0." The next problem is to specify a number of verbal phrases that could be used to express membership degrees in between. This is not a very difficult task but one that demands that all parties involved agree on what is meant by what. Thus, when we say "John is somewhat tall," we should agree that this means that John is tall to degree equal to, say, 0.50. Unfortunately, rules are made to be broken and so eventually people intentionally or unintentionally forget the correspondence between phrasal verbs and numbers. In order to avoid such problems, people working in fuzzy mathematics have defined four different methodologies to define membership degrees.

According to the *formalist* approach, we define mathematical functions that play the role of the membership function. Thus, these functions are used to assign membership degrees.

Example 3.1 Consider the set of real numbers and its fuzzy subset "real numbers close to 10." It has been proposed that this fuzzy subset can be define using the following function (see [83]):

$$R(x) = \frac{1}{1 + (x - 10)^2}.$$

The table that follows shows a few numbers and their corresponding membership degrees.

Number	Membership degree
0	0.010
2	0.015
4	0.027
6	0.059
8	0.200
10	1.000
12	0.200
14	0.059
16	0.027
18	0.015

Example 3.2 The global Multidimensional Poverty Index (MPI) is the result of an initiative to measure acute multidimensional poverty in at least 100 developing countries. The index is calculated by taking into account acute deprivations in health, education, and living standards that a man or a woman faces at the same time. Figure 3.2 shows the weights that are used to compute MPI.

In the Global Multidimensional Poverty Index 2023 [68, p. 4], the authors presented the case of Deepa, a woman that lives in a small island community in the hill tracts of Rangamati, Bangladesh. The text gives all the details concerning Deepa's situation which I will not reproduce here. However, according to the global Multidimensional Poverty Index (see Fig. 3.2), Deepa is poor. The reason is that her deprivation score, which is calculated by adding the various weights, is 0.444 ($\frac{1}{6} + 5 \times \frac{1}{18} = \frac{8}{18} \approx 0.444$).[3] Someone is not poor if his or her deprivation score is less that 0.333. Thus, we can say that someone whose deprivation score is less that 0.333 is poor to degree equal to 0. In order to define a fuzzy subset we have to map the interval [0.333, 1] to the interval [0, 1]. In general, when we want to map the interval $[a, b]$ onto the interval $[c, d]$, we have to use the following function

$$f(t) = c + \left(\frac{d - c}{b - a}\right)(t - a)$$

Thus, we can define the fuzzy subset of poor people using the recipe describe in Fig. 3.2 and then we have to "normalize" the results using the previous formula.

MPI values for a given country is the product of the proportion of people who live in poverty as defined above (H) times the average deprivation score among poor people (A). Thus, MPI $= H \times A$. The MPI ranges from 0 to 1, and higher values imply higher poverty. Using the MPI values for a set of countries, we can form the fuzzy subset of poor countries. ∎

Probability theory is a branch of mathematics that systematically studies the outcome of random events or *experiments* (e.g., which team is going to win UEFA's Champions League, the results of horse racing, the maximum and the minimum temperature in a city on s specific day, etc.). For any such random experiment we need to be aware of all possible outcomes and then to see whether the outcome of such an experiment is really random (i.e., we cannot predict it). Depending on the characteristics of the experiment it is possible to assign *probabilities* to the possible outcomes. These probabilities are numbers that belong to the unit interval. For example, when we throw a balanced, six-faced die the possible outcomes are 1, 2, 3, 4, 5, 6 and to each of them we assign a probability of 1/6. Thus, the probability that the next roll of the die will be 6 is 1/6. Some authors consider that probabilities and membership degrees are the same thing or at least two facets of the same concept. In fact, Zadeh believed this. However, this idea is totally wrong from a pragmatic and, more generally, a philosophic point of view. Suffice it to say that probabilities and

---

[3] The expression $a \approx b$ means that a is approximately equal to b.

Dimensions of Poverty	Indicator	Deprived if living in the household where…	Weight
Health	Nutrition	Any adult under 70 years of age or any child for whom there is nutritional information is under-nourished.	1/6
	Child mortality	Any child under the age of 18 years has died in the family in the five-year period preceding the survey	1/6
Education	Years of schooling	No household member aged "school entrance age" + six years or older has completed at least six years of schooling.	1/6
	School attendance	Any school-aged child is not attending school up to the age at which he/she would complete class eight.	1/6
Standard of living	Cooking Fuel	The household cooks with dung, wood, charcoal or coal.	1/18
	Sanitation	The household's sanitation facility is not im-proved (according to SDG guidelines) or it is im-proved but shared with other households.	1/18
	Drinking Water	The household does not have access to improved drinking water (according to SDG guidelines) or improved drinking water is at least a 30-minute walk from home, round trip.	1/18
	Electricity	The household has no electricity.	1/18
	Housing	At least one of the three housing materials for roof, walls and floor are inadequate: the floor is of natural materials and/or the roof and/or walls are of natural or rudimentary materials.	1/18

Fig. 3.2 The dimensions, indicators, deprivation cutoffs, and weights of the global Multidimensional Poverty Index

membership degrees measure completely different things. A probability is the likelihood degree of some event while a membership degree is a *truth* value (i.e., something that *asserts* to what degree something has a particular property). In order to make the difference as clear as possible, let me present an example borrowed from [8].

Example 3.3 Let D be the set of all potable liquids (i.e., liquids that are suitable for drinking). Moreover, suppose that you have not drink anything for three days (in the original example the time without a drink is one week but humans cannot stand more than 3–4 days without water). Suddenly, you are offered two bottles A and B and the person who offers these bottles tells you that bottle A belongs to D with a probability of 0.9 and bottle B has a degree of membership of 0.9 to D. From which of the bottles should you drink? A probability of 0.9 could mean that the bottle was chosen from a set of ten bottles in a fridge and nine of them contain water and one of them contains a liquid rodenticide. On the other hand, a membership degree equal to 0.9 means that the liquid is most probably drinkable (e.g., it may contain something whose best-before date has expired recently). I am more than sure that you would chose to drink from bottle B. ∎

In conclusion, it is totally wrong to use probabilities to define membership degrees.

A third approach to the problem of membership assignment is the one that uses *decision theory* [21]. This theory deals with cases where someone has to make a choice between several possible acts. The simplest case is the one when a person knows exactly the outcome of each act and, therefore, knows which act she has to chose. The most general case is the one where each act may lead to one of several outcomes, which depend on existing conditions. The theory assumes that each outcome depends on the act A and some state of the world ω, which are features of the world that the person has no control over.

Clearly, the person who makes the decisions has certain preferences between the various outcomes and between lotteries that have these outcomes as prizes. It is assumed that we know these preferences and we also know that these preferences make sense. It is possible to "mark" each outcome with a number that is called *utility*. In general, the decision maker has to choose the act that has the greatest expected utility. The theory assumes that there is a mechanism (i.e., a function) called *payoff function* that depends on A and ω that gives the utility of the outcome determined by any act A and world state ω.

Suppose now that we have the statement "Bob is smart." If we want to find the payoff function for this statement, we have to consider, for each possible "world state," the difference between the utility of the outcome that is, of what actually will happen, when the statement is assumed to be *true* or when it is assume to be *false*. Here true and false means that in general we agree that the statement holds or does not hold, respectively. Of course, it is quite complicated to see what is actually true of false, in the sense just described. For example, this depends on who makes this statement (e.g., if I will say that aliens will invade Earth most probably I will end in a madhouse, but if the president of the United States makes the same statement, then this means that they are already here) and other similar *details*.

Thus, it is a very complex way to give membership degrees using this method but it is quite *scientific*.

The fourth approach to the problem of membership assignment is the one that uses *axiomatic measurement theory* [38]. The idea behind this theory is that when we have to measure some attribute of, say, a group of people (i.e., their height), we associate a number (or, more generally, some mathematical entity) with the people in such a way that the properties of the attribute are represented by the numerical properties faithfully. For example, suppose we have the attribute height. Then, if "Bob is tall" and "Yannis is very tall," we can conclude that "Yannis is taller than Bob." Therefore, the statements "Bob is tall" and "Yannis is very tall" should be associated with the numbers a and b, respectively, and $a \leq b$. More formally, if we have an *empirical* ordering of some attribute, that is, the properties of the attribute are ordered in some way that can be verified or disproved by observation or experiment, and a function that maps these properties to real numbers in a way that preserves the ordering, then we have a mechanism to construct fuzzy subsets. The general theory and the particular mechanism by which we can assign membership degrees is much more involved (for example, the interested reader may check out [67, 73]), but the general idea has been just described. From now on I will assume that the membership degrees have been assigned using some meaningful method.

3.5   New Fuzzy Subsets from Old Ones

Assume that Isabella is a coin collector and that she has many coins. All of her coins are either copper nickel coins, silver coins, or copper plated steel coins. As expected, not all silver coins contain the same amount of pure silver. Thus, we need a way to express silver fineness. For this purpose we use the millesimal fineness (millesimal means one of 1000 equal parts). This is a number between 0 and 1000 that expresses the fineness of a precious metal object. For example, a coins that is made of fine silver has millesimal fineness equal to 999 and it contains 99.9% silver. Let us denote the set of Isabella's coins with C. Assume that A is the set of all silver coins regardless of their millesimal fineness and E the set of expensive coins, that is, all coins whose value is over 35 €. Then, if we want to find which silver coins are valuable, we have to find the set $A \cap E$, that is, the intersection of the sets. Similarly, if we want to find which coins are either silver or expensive, we have to find the set $A \cup E$. And if we want to find the cheap coins, we have to find the set E^C.

Consider again the set C and let us define two fuzzy subsets: The fuzzy subset of silver coins and the fuzzy set of valuable coins. We denote by S the fuzzy subset of silver coins and by V the fuzzy subset of valuable coins. If x is a silver coin with millesimal fineness equal to 850, then $S(x) = 0.85$. If y is the most expensive coin of Isabella's collection and its price is p €, and if the coin z costs q €, then $V(z) = q/p$. To the best of our knowledge, the purest silver coin ever produced had millesimal fineness equal to 999.99. Thus, there is no coin x in Isabella's collection for which $S(x) = 1$. This means that S is a subnormal fuzzy subset and V is a normal one. Having defined these fuzzy subsets, it is natural to ask what is the fuzzy subset of coins that are both silver and expensive? Obviously, we need to know the membership degrees of the elements of the this fuzzy subset. Similarly, we need to know what is the membership degree of the elements of the coins that are either silver of expensive. And of course we need to know to what degree a coin is cheap when we know that is expensive to degree equal to c.

The question posed in the previous paragraph can be answered by finding the intersection, the union, and the complement of the corresponding fuzzy subsets. Zadeh proposed to compute the new membership degrees by computing the minimum and the maximum of the membership degrees for the intersection and the union, correspondingly, and the operation $1 - d$ for the complement, where d is the membership degree of an element in the original fuzzy subset. His proposal was actually an extension of the definition of the ordinary set operations. Recall, that when we have two sets A and B, then their intersection consists of all these elements that belong to both sets. Let us see how this can be expressed using characteristic functions:

$$x \in A \cap B \text{ if and only if } \min(\chi_A(x), \chi_B(x)) = 1.$$

Similarly, the union of A and B that consists of all the elements of both A and B and this can be expressed using characteristic functions as follows:

$$x \in A \cup B \text{ if and only if } \max(\chi_A(x), \chi_B(x)) = 1.$$

Obviously, the condition $\max(\chi_A(x), \chi_B(x)) = 1$ is always true. The following mathematical statement defines the condition under which an element of the the universe of discourse belongs to the complement of A.

$$x \in A^C \text{ if and only if } 1 - \chi_A = 0.$$

Let us now define precisely the basic operations between fuzzy subsets. The reader should compare the following definitions with the previous ones.

Definition 3.5.1 Assume that $A : X \to [0, 1]$ and $B : X \to [0, 1]$ are two fuzzy subsets of X. Then,

- their *union* is the fuzzy subset

$$(A \cup B)(x) = \max(A(x), B(x));$$

- their *intersection* is the fuzzy subset

$$(A \cap B)(x) = \min(A(x), B(x));$$

- the *complement* of A is the fuzzy subset

$$A^C(x) = 1 - A(x), \quad \text{for all } x \in X.$$

Example 3.4 Consider the following class of students

$$\text{Class} = \{\text{Olivia, Amelia, Emma, Michael, Roger}\}$$

and the following two fuzzy subsets of set Class:

$$\text{competent} = \left\{ \frac{0.2}{\text{Olivia}}, \frac{0.5}{\text{Amelia}}, \frac{0.6}{\text{Emma}}, \frac{0.8}{\text{Michael}}, \frac{1.0}{\text{Roger}} \right\}$$

$$\text{clever} = \left\{ \frac{0.8}{\text{Olivia}}, \frac{0.6}{\text{Amelia}}, \frac{0.4}{\text{Emma}}, \frac{0.2}{\text{Michael}}, \frac{0.1}{\text{Roger}} \right\}.$$

Let us form the fuzzy subset of the students that are both competent and clever:

$$(\text{competent} \cap \text{clever})(\text{Olivia}) = \min(0.2, 0.8) = 0.2$$
$$(\text{competent} \cap \text{clever})(\text{Amelia}) = \min(0.5, 0.6) = 0.5$$
$$(\text{competent} \cap \text{clever})(\text{Emma}) = \min(0.6, 0.4) = 0.4$$
$$(\text{competent} \cap \text{clever})(\text{Michael}) = \min(0.8, 0.2) = 0.2$$
$$(\text{competent} \cap \text{clever})(\text{Roger}) = \min(1.0, 0.1) = 0.1.$$

This means that

$$\text{competent} \cap \text{clever} = \left\{ \frac{0.2}{\text{Olivia}}, \frac{0.5}{\text{Amelia}}, \frac{0.4}{\text{Emma}}, \frac{0.2}{\text{Michael}}, \frac{0.1}{\text{Roger}} \right\}.$$

Next, let us form the fuzzy subset of the students that are either competent or clever:

$$(\text{competent} \cup \text{clever})(\text{Olivia}) = \max(0.2, 0.8) = 0.8$$
$$(\text{competent} \cup \text{clever})(\text{Amelia}) = \max(0.5, 0.6) = 0.6$$
$$(\text{competent} \cup \text{clever})(\text{Emma}) = \max(0.6, 0.4) = 0.6$$
$$(\text{competent} \cup \text{clever})(\text{Michael}) = \max(0.8, 0.2) = 0.8$$
$$(\text{competent} \cup \text{clever})(\text{Roger}) = \max(1.0, 0.1) = 1.0.$$

Consequently,

$$\text{competent} \cup \text{clever} = \left\{ \frac{0.8}{\text{Olivia}}, \frac{0.6}{\text{Amelia}}, \frac{0.6}{\text{Emma}}, \frac{0.8}{\text{Michael}}, \frac{1.0}{\text{Roger}} \right\}.$$

Finally, let us compute the complement of the fuzzy subset clever.

$$\text{clever} = \left\{ \frac{0.8}{\text{Olivia}}, \frac{0.6}{\text{Amelia}}, \frac{0.4}{\text{Emma}}, \frac{0.2}{\text{Michael}}, \frac{0.1}{\text{Roger}} \right\}$$

$$\text{clever}^C = \left\{ \frac{0.2}{\text{Olivia}}, \frac{0.4}{\text{Amelia}}, \frac{0.6}{\text{Emma}}, \frac{0.8}{\text{Michael}}, \frac{0.0}{\text{Roger}} \right\}.$$

■

Equation 2.1 presents the basic properties of set complementation. Using the fuzzy subset competent defined above, let us see whether these properties do hold.

$$\text{competent} \cup \text{competent}^C = \left\{ \frac{0.8}{\text{Olivia}}, \frac{0.5}{\text{Amelia}}, \frac{0.6}{\text{Emma}}, \frac{0.8}{\text{Michael}}, \frac{1.0}{\text{Roger}} \right\} \neq \text{class}$$

$$\text{competent} \cap \text{competent}^C = \left\{ \frac{0.2}{\text{Olivia}}, \frac{0.5}{\text{Amelia}}, \frac{0.4}{\text{Emma}}, \frac{0.2}{\text{Michael}}, \frac{0.0}{\text{Roger}} \right\} \neq \varnothing_{\text{Class}}$$

Here $\varnothing_{\text{Class}}$ is the *empty* fuzzy set, that is, the set fuzzy subset of class for which the following property holds true:

$$\text{for all } x \in \text{Class: } \varnothing_{\text{Class}}(x) = 0.$$

Also, class is the characteristic function of Class. The property $A \cap A^C = \varnothing$ is called the *law of contraction* and the property $A \cup A^C = U$ is the law of the excluded middle. This means that these two basic laws of set theory do not hold in fuzzy set theory since we have found at least one fuzzy subset for which these laws do not hold.

On the other hand, the De Morgan laws (see Eq. 2.2) do hold for fuzzy sets. Suppose we want to prove that $(A \cup B)^C(x) = A^C(x) \cap B^C(x)$ for all $x \in X$. We note that the membership degree of $(A \cup B)^C(x)$ is

$$1 - \max\bigl(A(x), B(x)\bigr) = 1 + \min\bigl(-A(x), -B(x)\bigr)$$
$$= \min\bigl(1 - A(X), 1 - B(x)\bigr),$$

which is the membership degree of $A^C \cap B^C$. Similarly, we can prove the validity of the second rule.

There are a few more operation between fuzzy set that are presented in the definition that follows.

Definition 3.5.2 Let $A : X \rightarrow [0, 1]$ and $B : X \rightarrow [0, 1]$ be two fuzzy subsets of X. Then,

- their algebraic product is
$$(AB)(x) = A(x) \cdot B(x);$$

- A is a *subset* of B, denoted by $A \subseteq B$, if and only if

$$A(x) \leq B(x), \text{ for all } x \in X \text{ and}$$

- the fuzzy *powerset* of X (i.e., the set of all ordinary fuzzy subsets of X) is denoted by $\mathscr{F}(X)$.

3.6 Triangular Norms and Conorms

Suppose again that we have a class of students and that we create the fuzzy subsets of beautiful or handsome students and the fuzzy subset of good students. If Angeline is beautiful to degree 0.75 and a good student to degree 0.50, then, according to what we know already, she is both a beautiful and a good student to degree 0.50. Although, this is *correct* according to the theory of fuzzy subsets, still some people will disagree and expect a higher or even a lower value for this degree. For instance, it has been shown that attractive students may get better grades when they attend in-person classes [43]. In particular, it has been observed that this is more evident for students attending classes of non-quantitative subjects (i.e., subjects where students have group assignments, seminars, and oral presentations and a final exam). On the other hand, attractiveness has no impact for students attending classes of quantitative subjects (i.e., subjects where students have to give only a final written exam, e.g., physics or mathematics classes). Thus, we can say that Angeline is both a beautiful and a good student to degree 0.70!

The previous discussion proves that the standard way to compute the membership degrees of the intersection and the union of two fuzzy subsets is not adequate in, at least, some cases. Is it possible to solve this problem? In different words, is it possible to find other means to compute the membership degrees for the these two operations? The answer to this question is the use of *triangular norms* or just *t-norm* and *triangular conorms* or just *t-conorms*. These are functions that were introduced by Berthold Schweizer and Abe Sklar, following some ideas of Karl Menger (see [52] for a historical overview of t-norms and t-conorms). But what is so special about these functions? In a nutshell, t-norms and t-conorms have all the characteristics of min and max, respectively. The following definition makes this precise for t-norms.

Definition 3.6.1 A t-norm is a function with two arguments that belong to the unit interval and yields a number that also belongs to the unit interval. In addition, a t-norm T should have the following properties:

(T1) $T(x, y) = T(y, x)$,
(T2) $T(x, T(y, z)) = T(T(x, y), z)$,
(T3) $T(x, y) \leq T(x, z)$, when $y \leq z$, and
(T4) $T(x, 1) = x$.

It is not difficult to see that min is a t-norm. First we note that $min(x, y) = x$ if $x \leq y$, else $min(x, y) = y$. Using this definition, we see that if $x \leq y$, then $\min(x, y) = x$ and $\min(y, x) = x$, therefore $\min(x, y) = \min(y, x)$. Similarly, we can show the second property when $T = \min$. Let us prove the third property. Suppose that $x \leq y$ and $x \leq z$, then $\min(x, y) = x$ and $\min(x, z) = x$ and obviously $x \leq x$. Now, assume that $x \geq y$ and $x \geq z$, then $\min(x, y) = y$ and $\min(x, z) = z$ and we know that $y \leq z$. Next, assume that $x \leq y$ and $x \geq z$, but this is not possible if $y \leq z$. Finally, assume that $x \geq y$ and $x \leq z$, then $\min(x, y) = y$ and $min(x, z) = x$, but we already know that $y \leq x$. The fourth property is true because for any $x \in [0, 1]$, we know that $x \leq 1$ and so $\min(x, 1) = x$.

In many cases, we want t-norms to have the form of binary operand (i.e., to operate like the symbols $+$, $\times$, etc.) and then the previous properties take the following form when $*$ denotes a t-norm.

(T1) $x * y = y * x$,
(T2) $x * (y * z) = (x * y) * z$,
(T3) $x * y \leq x * z$, when $y \leq z$, and
(T4) $x * 1 = x$ and $x * 0 = 0$.

There are many t-norms and below I present some of them.

Minimum t-norm This is the usual minimum function which is called the Gödel t-norm. Compared to any other t-norm, min gives always the largest value, that is, for any t-norm T, it holds that $T(x, y) \leq \min(x, y)$ for all $x, y \in [0, 1]$.

Product t-norm The product of two numbers is a t-norm.

Łukasiewicz t-norm This t-norm is defined as follows: $T(x, y) = \max(0, x + y - 1)$.

Drastic t-norm This t-norm is a piecewise-defined function, that is, a function that behaves differently based on its argument or arguments. The drastic t-norm is defined as follows.

$$T(x, y) = \begin{cases} y & \text{if } x = 1 \\ x & \text{if } y = 1 \\ 0 & \text{otherwise.} \end{cases}$$

Nilpotent minimum This t-norm is also a piecewise-defined function:

$$T(x, y) = \begin{cases} \min(x, y) & \text{if } x + y > 1 \\ 0 & \text{otherwise.} \end{cases}$$

Hamacher product One more example of a t-norm defined by a piecewise function:

$$T(x, y) = \begin{cases} 0 & \text{if } x = y = 0 \\ \frac{ab}{a+b-ab} & \text{otherwise.} \end{cases}$$

The definition of t-conorms is similar to the definition of t-norms:

Definition 3.6.2 A t-conorm is a function with two arguments that belong to the unit interval and yields a number that also belongs to the unit interval. In addition, a t-norm S should have the following properties:

(S1) $S(x, y) = S(y, x)$,
(S2) $S(x, S(y, z)) = S(S(x, y), z)$,
(S3) $S(x, y) \leq S(x, z)$, when $y \leq z$, and
(S4) $S(x, 0) = x$.

It is not difficult to prove that max is indeed a t-conorm and the reader is invited to try to prove it using arguments similar to the proof that min is a t-norm.

When a t-norm has the form of binary operator, then we want the t-conorm to be also a binary operator (usually denoted by the symbol $\star$) and the properties of the previous definition take the form that follows.

(S1) $x \star y = y \star x$,
(S2) $x \star (y \star z) = (x \star y) \star z$,
(S3) $x \star y \leq x \star z$, when $y \leq z$, and
(S4) $x \star 0 = x$ and $x \star 1 = 1$.

Given a t-norm T, its dual t-conorm S is defined by the following rule

$$S(x, y = 1 - T(1 - x, 1 - b).$$

The duals of the t-norms presented above are:

Maximum t-conorm This is the usual maximum function. Compared to any other t-conorm, max gives always the smallest value, that is, for any t-conorm S, it holds that $\max(x, y) \leq S(x, y)$ for all $x, y \in [0, 1]$.
Probabilistic sum Is dual to the product t-norm and is defined by $S(x, y) = x + y - x \cdot y$.
Bounded sum Is dual to the Łukasiewicz t-norm and is defined by $S(x, y) = \min(x + y, 1)$.
Drastic t-conorm Is dual to the drastic t-norm and is defined by

$$S(x, y) = \begin{cases} y & \text{if } x = 0 \\ x & \text{if } y = 0 \\ 1 & \text{otherwise.} \end{cases}$$

Nilpotent maximum Is dual to the nilpotent minimum and is defined by

$$S(x, y) = \begin{cases} \max(x, y) & \text{if } x + y < 1 \\ 1 & \text{otherwise.} \end{cases}$$

Einstein sum Is a dual to one of the Hamacher t-norms and is defined by

$$S(x, y) = \frac{x + y}{1 + x \cdot y}.$$

Choosing the proper t-norm and its dual t-conorm is not an easy task. However, some researchers have investigated this problem to some extend (e.g., see [1] for a discussion of this problem and its solution in a specific context). However, in some cases there is a set of observed values (e.g., values produced by experts that may use certain decision patterns) and it is necessary to define a t-norm that should match some of the observed values. Of course, one should not expect error-free observations and this is why the problem is called an *approximation* problem. Such problems belong to a general category of nonlinear least-squares (nonlinear data-fitting) problems and they can be solved with a number of methods (e.g., see [7]).

3.7 "Intuitionistic" Fuzzy Subsets

In Sect. 1.1 I talked about truth values and presented the idea that we can create additional truth values by assuming that 0 denotes falsehood, a natural number n denotes truth, and the numbers in between denote partial truth or falsity, if you prefer this. However, it is quite possible to assume that again 0 denotes falsehood, 1 denotes truth and all the rational numbers in between denote the additional truth values. In this case, it is easy to see that the truth values are infinite! In this setting, when we say that something is true to degree equal to d, then, usually, it is false to degree equal to $1 - d$. This seems quite reasonable since the *negation* corresponds to set complementation. But, is it possible to have another way to get the complement of a fuzzy subset or the truth degree of the negation of a statement?

It is quite possible to use other functions instead of the function $N(x) = 1 - x$ to compute the negation or the membership degree of each element of the complement of a fuzzy subset. However, if N is a function that is supposed to realize the complementation operation, then it should have at least the following two properties:

(N1) $N(0) = 1$ and $N(1) = 0$.
(N2) For all $x, y \in [0, 1]$, if $x \leq y$, then $N(x) \geq N(y)$.

In many cases of practical significance, it is expected that a fuzzy complementation operation should be *involutive*, that is,

(N3) $N(N(x)) = x$ for each $x \in [0, 1]$.

Example 3.5 The following function realizes a fuzzy complementation operation which is not involutive:

$$N(x) = \frac{1}{2}(1 + \cos \pi x).$$

On the other hand, the following class of functions

$$N_\lambda(x) = \frac{1-x}{1+\lambda x},$$

where λ is either a positive real number or equal to -1, defines an involutive fuzzy complementation operation for each value of the parameter λ.   ∎

Although the use of alternative negation operators is something quite meaningful, still Krassimir Atanassov (Красимир Атанасов) thought it would be better to incorporate the nonmembership degree of any element in the definition of a fuzzy set. This choice is the basis of his *"intuitionistic"*[4] fuzzy subsets [2–4]. Let us see what really is an "intuitionistic" fuzzy subset.

Definition 3.7.1 Suppose that E is an ordinary set and $A \subset E$. Then, an "intuitionistic" fuzzy subset (IFS) A of E is a set of triples of the following form

$$A = \big\{\langle x, \mu_A(x), \nu_A(x)\rangle \mid x \in E\big\},$$

where $\mu_A : E \rightarrow [0, 1]$ and $\nu_A : E \rightarrow [0, 1]$ are two functions that define the membership and the nonmembership degrees of any element $x \in A$. In addition, the membership and the nonmembership degrees must satisfy the following condition

$$0 \leq \mu_A(x) + \nu_A(x) \leq 1.$$

Thus, the main difference between an ordinary fuzzy subset and an IFS is that for the definition of the later we need to specify two functions whereas for the definition of the former we need only one function. In addition, the reader should note that Atanassov has practically defined an IFS as a set that consists of triples. In different words, he defined IFS as ordinary sets and this has some unexpected consequences as we will see in what follows.

Example 3.6 Suppose that E is the set of all countries with elective governments. After each election battle in $c \in E$, we know the percentage of the electorate who have voted for the corresponding government. I will denote this percentage by $m(x)$. Then, we can define $\mu(x) = m(x)/100$. For ordinary fuzzy sets, $\nu(x) = 1 - \mu(x)$ and this number corresponds

---

[4] The term was proposed by George Gargov, nevertheless, many researchers in the fuzzy set community objected to the use of this term (e.g., see [18]). However, I feel this paper is a fierce polemic against Atanassov's work. To avoid confusion with mathematical intuitionism, I will use the term intuitionistic in quotes.

to the votes that were totally against the government. However, if we exclude invalid and blank ballot papers and count only the votes against the government, then $v(x) \neq 1 - \mu(x)$ and we can use this number to construct this IFS. This example is due to Adhiyaman Manickam.   ■

If you wonder how, in general, one could define the nonmembership degrees, then the answer is simple: Using the techniques and the tools described in Sect. 3.4. From the previous definition it should be obvious that any ordinary fuzzy subset is actually an IFS:

$$\{\langle x, A(x), 1 - A(x)\rangle \mid x \in E\},$$

where A is fuzzy subset. In addition, we can define the following function

$$\pi_A(x) = 1 - \mu_A(x) - v_A(x), \tag{3.1}$$

that defines the degree of *nondeterminacy* (uncertainty) of the membership of an element $x \in E$ to A. Ronald Robert Yager [75] called this quantity the *hesitancy* of x and it is considered to be lack of commitment or uncertainty associated with the membership degree of x. In more conventional words consider the case of an election for a public office. A candidate will get a number of votes (this can be modeled by the membership degree), voters will cast their vote for other candidates (this can be modeled by the nonmembership degree), and, as expected some voters will not vote or they will cast an invalid ballot paper (this can be modeled by the hesitancy degree). Note that in the case of ordinary fuzzy subsets, $\pi_A(x) = 0$ for all $x \in E$. Now I will present the definition of basic relations and operations on IFSs.

Definition 3.7.2 Assume that $\{\langle x, \mu_A(x), v_A(x)\rangle \mid x \in X\}$ and $\{\langle x, \mu_B(x), v_B(x)\rangle \mid x \in X\}$ are two IFSs A and B. Then,

- $A \subset B$ if and only if for all $x \in X$

$$\mu_A(x) < \mu_B(x) \quad \text{and} \quad v_A(x) > v_B(x);$$

- $A \subseteq B$ if and only if for all $x \in X$

$$\mu_A(x) \leq \mu_B(x) \quad \text{and} \quad v_A(x) \geq v_B(x);$$

- $A = B$ if and only if
$$A \subset B \quad \text{and} \quad B \subset A;$$

- their *union* is

$$A \cup B = \left\{\left\langle x, \max[\mu_A(x), \mu_B(x)], \min[v_A(x), v_B(x)]\right\rangle \,\middle|\, x \in X\right\};$$

- their *intersection* is

$$A \cap B = \left\{ \left\langle x, \min[\mu_A(x), \mu_B(x)], \max[\nu_A(x), \nu_B(x)] \right\rangle \mid x \in X \right\};$$

- the *complement* of A is the IFS A^C

$$A^C = \left\{ \langle x, \nu_A(x), \mu_A(x) \rangle \mid x \in X \right\};$$

- their *difference* is

$$A \setminus B = \left\{ \left\langle x, \min[\mu_A(x), \nu_B(x)], \max[\nu_A(x), \mu_B(x)] \right\rangle \mid x \in X \right\};$$

- their *symmetric difference* is

$$A \triangle B = \left\{ \left\langle x, \min[\max(\mu_A(x), \mu_B(x)), \max(\nu_A(x), \nu_B(x))], \right. \right.$$
$$\left. \left. \max[\min(\nu_A(x), \nu_B(x)), \min(\mu_A(x), \mu_B(x))] \right\rangle \mid x \in X \right\};$$

- their *sum* is

$$A + B = \left\{ \left\langle x, \mu_A(x) + \mu_B(x) - \mu_A(x) \cdot \mu_B(x), \nu_A(x) \cdot \nu_B(x) \right\rangle \mid x \in X \right\};$$

- their *product* is

$$A \cdot B = \left\{ \left\langle x, \mu_A(x) \cdot \mu_B(x), \nu_A(x) + \nu_B(x) - \nu_A(x) \cdot \nu_B(x) \right\rangle \mid x \in X \right\};$$

- the *necessity* of A is the IFS $\square A$

$$\square A = \left\{ \langle x, \mu_A(x), 1 - \mu_A(x) \rangle \mid x \in X \right\};$$

- the *possibility* of A is the IFS $\Diamond A$

$$\Diamond A = \left\{ \langle x, 1 - \nu_A(x), \mu_A(x) \rangle \mid x \in X \right\}.$$

Atanassov has claimed "[f]ollowing the idea of a fuzzy set of level α [...] the definition of a set of (α, β)-level, generated by an IFS A." Essentially, he claimed that the (α, β)-level sets are an extension of the notion of α-cut for fuzzy sets. However, as the definition, that I will present in a moment, shows, that the two notions are entirely different since the α-cut of a fuzzy set is a crisp set while the (α, β)-level set of an IFS is another IFS!

Definition 3.7.3 If $A = \{\langle x, \mu_A(x), \nu_A(x) \rangle \mid x \in X\}$ is an IFS and $\alpha, \beta \in [0, 1]$, such that $\alpha + \beta \leq 1$, then the (α, β)-level set of A is the following IFS:

$$N_{\alpha,\beta}(A) = \{\langle x, \mu_A(x), \nu_A(x)\rangle \mid x \in X \text{ and } \mu_A(x) \geq \alpha \text{ and } \nu_A(x) \leq \beta\}.$$

Remember that it was noted above that the definition of IFSs has some unexpected consequences. The following IFS is called *the set of level of membership* α generated by A.

$$N_\alpha(A) = \{\langle x, \mu_A(x), \nu_A(x)\rangle \mid x \in X \text{ and } \mu_A(x) \geq \alpha\}$$

Also, the following IFS is called *the set of level of nonmembership* α generated by A.

$$N^\alpha(A) = \{\langle x, \mu_A(x), \nu_A(x)\rangle \mid x \in X \text{ and } \nu_A(x) \leq \alpha\}$$

Atanassov and his collaborators have introduced many other operators but I feel most of them are just mathematical curiosities and do not provide a better insight in the understanding of vagueness. Thus, I will say nothing more about IFSs.

Example 3.7 Consider the following class of students that we used in the previous section

$$\text{Class} = \{\text{Olivia, Amelia, Emma, Roger}\}$$

and the following two IFSs of set Class:

competent $= \{\langle\text{Olivia}, 0.2, 0.7\rangle, \langle\text{Amelia}, 0.5, 0.4\rangle \langle\text{Emma}, 0.6, 0.4\rangle, \langle\text{Roger}, 0.1, 0.8\rangle\}$
clever $= \{\langle\text{Olivia}, 0.8, 0.1\rangle, \langle\text{Amelia}, 0.6, 0.3\rangle \langle\text{Emma}, 0.4, 0.5\rangle, \langle\text{Roger}, 0.1, 0.5\rangle\}$

Let us form the intersection of these two IFSs:

$$\begin{aligned}
(\text{competent} \cap \text{clever})(\text{Olivia}) &= \langle\min(0.2, 0.8), \max(0.7, 0.1)\rangle = \langle 0.2, 0.7\rangle \\
(\text{competent} \cap \text{clever})(\text{Amelia}) &= \langle\min(0.5, 0.6), \max(0.4, 0.3)\rangle = \langle 0.5, 0.4\rangle \\
(\text{competent} \cap \text{clever})(\text{Emma}) &= \langle\min(0.6, 0.4), \max(0.4, 0.5)\rangle = \langle 0.4, 0.5\rangle \\
(\text{competent} \cap \text{clever})(\text{Roger}) &= \langle\min(1.0, 0.1), \max(0.8, 0.5)\rangle = \langle 0.1, 0.8\rangle
\end{aligned}$$

This means that

competent$\cap$clever $=$
$$\{\langle\text{Olivia}, 0.2, 0.7\rangle, \langle\text{Amelia}, 0.5, 0.4\rangle, \langle\text{Emma}, 0.4, 0.5\rangle, \langle\text{Roger}, 0.1, 0.8\rangle\}.$$

Next, let us form the fuzzy subset of the students that are either competent or clever:

$$\begin{aligned}
(\text{competent} \cup \text{clever})(\text{Olivia}) &= \langle\max(0.2, 0.8), \min(0.7, 0.1)\rangle = \langle 0.8, 0.1\rangle \\
(\text{competent} \cup \text{clever})(\text{Amelia}) &= \langle\max(0.5, 0.6), \min(0.4, 0.3)\rangle = \langle 0.6, 0.3\rangle \\
(\text{competent} \cup \text{clever})(\text{Emma}) &= \langle\max(0.6, 0.4), \min(0.4, 0.5)\rangle = \langle 0.6, 0.4\rangle \\
(\text{competent} \cup \text{clever})(\text{Roger}) &= \langle\max(1.0, 0.1), \min(0.8, 0.5)\rangle = \langle 1.0, 0.5\rangle
\end{aligned}$$

Consequently,

competent $\cup$ clever $=$

$$\{\langle Olivia, 0.8, 0.1\rangle, \langle Amelia, 0.6, 0.3\rangle, \langle Emma, 0.6, 0.4\rangle, \langle Roger, 1.0, 0.5\rangle\} .$$

■

Example 3.8 Suppose we have a class of students and we want to examine the social status of each student within her class. Zuzana Heinzová and Ján Belko[28] have proposed a simple method that can be used to examine answers to a questionnaire very carefully in order to discover information about the social status of students. In particular, after asking every student to rate each of his classmates with a positive or a negative vote, they proposed the following indexes

$$I_p(x) = \frac{\text{positive votes for student } x}{\text{negative votes for student } x} \text{ and } I_n(x) = -\frac{\text{negative votes for student } x}{\text{positive votes for student } x},$$

as a means to measure the social status of student x. Naturally, it is quite possible that some students have no negative votes or no positive votes. Magdaléna Renčová [50] used these ideas to form an IFS that would describe the popularity of each student. To overcome the problem of zero votes, she proposed the use of different indexes:

$$I'_p(x) = \frac{\text{positive votes for student } x}{N} \text{ and } I'_n(x) = \frac{\text{negative votes for student } x}{N},$$

where N is the total number of students. Next, she used these new indexes to form the IFS

$$S = \{\langle x, I'_p(x), I'_n(x)\rangle \mid x \in X\}.$$

The general acceptance index becomes $I_a(x) = I'_p(x) - I'_n(x)$.

Piotr Dworniczak spotted a problem in Renčová's definition. In particular, if a student has no positive and no negative votes and another one has $N/2$ positive votes and $N/2$ negative votes, then both have the same acceptance index. Obviously, this result does not reflect reality. Thus, he proposed a more elaborate method to grade the popularity of a student. Instead of a "yes" or "no," he proposed the following choice of answers:

(1) strongly accepted,
(2) accepted,
(3) rather accepted,
(4) rather not accepted,
(5) not accepted,
(6) definitely not accepted,
(7) I cannot determine whether I accept this person or not. (I don't now).

Then, he proposed to grade these responses as follows:

Linguistic evaluation given by x_i for classmate x_j	$(\mu_{i,j}, \nu_{i,j})$
Strongly Accepted	$(1, 0)$
Accepted	$(0.66, 0)$
Rather Accepted	$(0.33, 0)$
Rather Not Accepted	$(0, 0.33)$
Not Accepted	$(0, 0.66)$
Definitely Not Accepted	$(0, 1)$
I Don't Now	$(0, 0)$

From these he defined the acceptance index and the non-acceptance index for each student x_j as follows:

$$I'_p(x_j) = \frac{\mu_{1,j} + \cdots + \mu_{j-1,j} + \mu_{j+1,j} + \cdots + \mu_{N,j}}{N - 1}$$

and

$$I'_n(x_j) = \frac{\nu_{1,j} + \cdots + \nu_{j-1,j} + \nu_{j+1,j} + \cdots + \nu_{N,j}}{N - 1}.$$

Using these values we can form an IFS that describes the acceptance index of each student. In addition, the hesitancy provides information about the clarity of evaluation of the j-th person. ∎

3.8   Pythagorean Fuzzy Subsets

Although IFSs can be used to handle many cases, still there are cases that cannot be handled by them. For example, consider the case where the membership degree for some element x is $\frac{\sqrt{3}}{2}$ and the nonmembership degree is $\frac{1}{2}$, then

$$\frac{\sqrt{3}}{2} + \frac{1}{2} \approx 0.866 + 0.5 > 1.$$

Obviously, the theory of IFSs does not allow these numbers to be membership and non-membership degrees. However, one should notice that for $a, b \in [0, 1]$, $a^2 \leq a$ and $b^2 \leq b$. Therefore, if $a + b \leq 1$, then clearly $a^2 + b^2 \leq 1$. Thus, if a and b denote the membership and nonmembership degrees of an IFS, we can use the condition $a^2 + b^2 \leq 1$ to define an extension of IFSs:

Definition 3.8.1 Suppose that E is an ordinary set and $A \subset E$. Then, an *Pythagorean* fuzzy subset (PFS) A of E is a set of triples of the following form

$$A = \{\langle x, \mu_A(x), \nu_A(x) \rangle \mid x \in E\},$$

where $\mu_A : E \rightarrow [0, 1]$ and $\nu_A : E \rightarrow [0, 1]$ are two functions that define the membership and the nonmembership degrees of any element $x \in A$. In addition, the membership and the nonmembership degrees must satisfy the following condition

$$0 \leq (\mu_A(x))^2 + (\nu_A(x))^2 \leq 1$$

From this definition we can easily deduce that all IFSs are PFSs but not all PFSs are IFSs. A direct consequence of this remark is that the operations between any two PFS A and B are exactly the operations of the corresponding IFSs. There are many extensions of PFSs (e.g., see [47]) but in my opinion most of them are just mathematical curiosities.

3.9 Neutrosophic Fuzzy Subsets

As I explained above, for any IFS A, Eq. 3.1 defines the nondeterminacy of the membership of an element $x \in E$ to A. Obviously, here we calculate the nondeterminacy by using a formula. By following the ideas of Atanassov, Florentin Smarandache [54] introduced his *neutrosophic*[5] fuzzy subsets and made nondeterminacy or *indeterminacy*, as he call it, a third component of their definition, just like Atanassov made non-membership the second components of his IFS. Smarandache's original definition of his neutrosophic sets is quite involved and not so useful. Thus, he and his colleagues proposed a simplified version of his neutrosophic sets that are called single valued neutrosophic sets [48]:

Definition 3.9.1 Assume that X is some universe. Then, a single valued neutrosophic set (SVNS) A is characterized by a membership, a indeterminacy, and a nonmembership function. These functions are denoted by T_A, I_A, and F_A, respectively. For each $x \in X$: $T_A(x), I_A(x), F_A(x) \in [0, 1]$. Thus, a SNVS A can be written as

$$A = \{\langle x, T_A(x), I_A(x), F_A(x) \rangle \mid x \in X\}$$

where $0 \leq T_A(x) + I_A(x) + F_A(x) \leq 3$.

The term *neutrosophy* derives from the French word *neutre* that, in turn, derives from the Latin *neuter*, which means neutral, and from the Greek word *sophia* (σοφία), which means wisdom. Thus, the term neutrosophic means knowledge of neutral thought.

Since many people had difficulty to understand what this indeterminacy degree, Smarandache wrote an article [53] where he tried to explain with simple words the meaning of

---

[5] Contra to what Atanassov did, Smarandache coined a new term that did not cause confusion in the world of mathematics of vagueness.

this degree. According to him, given some attribute, the membership degree is about the degree that an element has this attribute (e.g., being tall), the nonmembership degree is the degree to which the element does not have this attribute (e.g., not being tall but short), and the indeterminacy degree expresses the degree to which an element has an attribute in the middle (e.g., having an average height). Here are some triples of attributes that express this idea:

Membership	Indeterminacy	Nonmembership
Male	Transgender	Female
Small	Medium	Large
Thin	Normal weight	Fat

When we have a fuzzy set $A : X \rightarrow [0, 1]$, then the membership degree of an element x is equal to $A(x)$, it nonmembership degree is equal to $1 - A(x)$ and its indeterminacy degree is equal to 0. In the case of an IFS A, the membership degree is equal to $\mu_A(x)$, the nonmembership degree is equal to $\nu_A(x)$, and the indeterminacy degree is equal to $1 - \mu_A(x) - \nu_A(x)$.

As expected, the various operations between SVNSs are extensions of the corresponding operations of IFS [16] and in what follows I give the definitions of these operations.

Definition 3.9.2 Assume that

$$A = \{\langle x, T_A(x), I_A(x), F_A(x)\rangle \mid x \in X\}$$

and

$$B = \{\langle x, T_B(x), I_B(x), F_B(x)\rangle \mid x \in X\}$$

are two SVNSs. Then,

- $A \subseteq B$ if and only if for all $x \in X$

$$T_A(x) \leq T_B(x), I_A(x) \leq I_B(x) \text{ and } F_A(x) \geq F_B(x);$$

- $A \subset B$ if and only if for all $x \in X$

$$T_A(x) < T_B(x), I_A(x) < I_B(x) \text{ and } F_A(x) > F_B(x);$$

- $A = B$ if and only if

$$A \subset B \quad \text{and} \quad B \subset A;$$

- their *union* is

$$A \cup B = \left\{ \left\langle x, \max[T_A(x), T_B(x)], \max[I_A(x), I_B(x)], \min[F_A(x), F_B(x)] \right\rangle \mid x \in X \right\};$$

- their *intersection* is

$$A \cap B = \left\{ \left\langle x, \min[T_A(x), T_B(x)], \min[I_A(x), I_B(x)], \max[F_A(x), F_B(x)] \right\rangle \mid x \in X \right\};$$

- the *complement* of A is the SVNS A^C

$$A^C = \left\{ \langle x, F_A(x), 1 - I_A(x), T_A(x) \rangle \mid x \in X \right\};$$

- their *difference* is

$$A \setminus B = \left\{ \left\langle x, \min[T_A(x), T_B(x)], \min[I_A(x), 1 - I_B(x)], \max[F_A(x), F_B(x)] \right\rangle \mid x \in X \right\}.$$

The α-cut of an SVNS is defined below.

Definition 3.9.3 Assume that

$$A = \{ \langle x, T_A(x), I_A(x), F_A(x) \rangle \mid x \in X \}$$

is a single-valued neutrosophic set of the universe X. For $\alpha \in [0, 1]$, the α-cut of A is the crisp set $^{\alpha+}A$ defined as follows:

$$^{\alpha+}A = \left\{ x \mid x \in X \text{ and either } T_A(x) \geq \alpha \text{ and } I_A(x) \geq \alpha \text{ or } F_A(x) \leq 1 - \alpha \right\}.$$

3.10   Bipolar Fuzzy Subsets

Bipolarity is widely understood to be the requirement that statements must be capable of being true and capable of being false [24]. But what does this requirement mean? In case we (wrongly) assume that statements can be either true or false, then they have something to do with what Ludwig Josef Johann Wittgenstein called *sense*. This is determined by the two poles: True and false. However, all this can be simplified if we assume that there are three truth values—false, half-true, and true [66]. These truth values correspond to negative, neutral, and positive. And I am sure that if Wittgenstein had thought of this analogy, then maybe his analysis would look far-fetched for his time, but still it would be very pioneering. An interesting idea related to bipolarity was put forth by Arturo Tozzi [66]: bipolarity is strongly related to the mathematical treatment of *paraconsistency*, which is the idea that a statement (e.g., John is tall) and its negation (e.g., John is not tall) are both true. In addition, a *paraconsistent logic* is a mathematical tool to reason about inconsistent information without yielding absurd statements. According to Tozzi, we can assume that the positive pole is a true statement and the negative pole is the negation of this statement, which is necessarily true. Although, this sounds a bit strange, the idea of a logic of paraconsistency is very old:

It seems that Nicholas of Autrecourt (c. 1300–1369) was the first philosopher to discuss it thoroughly in his *Universal Treatise* [5]. What is even more interesting is that one can use paraconsistency to "handle" vagueness. Basically, the idea is when we have a statement of the form "this t-shirt is red" and its negation "this t-shirt is not red," then both statements are true since we are not really sure about the color of this particular t-shirt. This idea is called *paraconsistent vagueness* [31].

As is usually the case, it was just a matter of time for someone to propose the idea of a fuzzy version of bipolarity. Indeed, Wen-Ran Zhang [82] proposed what he called *bipolar fuzzy sets*. Unfortunately, Zhang's paper is highly unreadable and does not give explicit definition of bipolar fuzzy sets. Fortunately, Jeong-Gon Lee and Kul Hur [42] have elaborated on this matter and our presentation is based on their work. After this rather long introduction, it is time to see what is a bipolar fuzzy subset.

Definition 3.10.1 Assume that X is a non-empty set. Then, the pair $A = (A^-, A^+)$ is called a bipolar fuzzy subset in X, if $A^+ : X \to [0, 1]$ and $A^- : X \to [-1, 0]$ are functions.

The empty bipolar fuzzy subset and the whole bipolar fuzzy subset are the bipolar fuzzy subsets $\mathbf{0}_{bp} = (\mathbf{0}_{bp}^-, \mathbf{0}_{bp}^+)$ and $\mathbf{1}_{bp} = (\mathbf{1}_{bp}^-, \mathbf{1}_{bp}^+)$, respectively, that have the following properties

$$\mathbf{0}_{bp}^-(x) = 0 \text{ and } \mathbf{0}_{bp}^-(x) = 0 \text{ for all } x \in X,$$

and

$$\mathbf{1}_{bp}^-(x) = 1 \text{ and } \mathbf{1}_{bp}^-(x) = -1 \text{ for all } x \in X.$$

Let $F = (F^-, F^+)$ be a bipolar fuzzy subset that describes some property (e.g., whether some person x is a friend or an enemy of a specific person, say Jane). Then, $F^+(x)$ is the degree to which x satisfies the property (e.g., the degree to which x is a friend of Jane) and $F^-(x)$ is the degree to which x satisfies the opposite property (e.g., the degree to which x is an enemy of Jane). If $F^+(x) = 0$, then x does not have the property (e.g., x is not a friend of Jane). Similarly, if $F^-(x) = 0$, then x does not have the opposite property (e.g., x is not an enemy of Jane).

Definition 3.10.2 Assume that $A = (A^-, A^+)$ and $B = (B^-, B^+)$ are two bipolar fuzzy subsets. Then,

- $A \subset B$ if and only if for all $x \in X$

$$A^+(x) < B^+(x) \text{ and } A^-(x) > B^-(x);$$

- $A \subseteq B$ if and only if for all $x \in X$

$$A^+(x) \leq B^+(x) \text{ and } A^-(x) \geq B^-(x);$$

- $A = B$ if and only if

$$A \subset B \quad \text{and} \quad B \subset A;$$

- the *complement* of A is the bipolar fuzzy set A^C

$$A^C(x) = \left(-1 - A^+(x), 1 - A^+(x)\right);$$

- their *union* is

$$(A \cup B)(x) = \left(\min\left(A^-(x), B^-(x)\right), \max\left(A^+(x), B^+(x)\right)\right);$$

- their *intersection* is

$$(A \cap B)(x) = \left(\max\left(A^-(x), B^-(x)\right), \min\left(A^+(x), B^+(x)\right)\right);$$

3.11   Interval Valued Fuzzy Subsets

When we ask someone to what degree Amelia is tall and he replies with, say, 0.78, then one may argue that we are talking about a very *precise* vagueness. In different words, how can be so sure that this number expresses the height of Amelia? Of course the answer is: We are not really sure, but we just give an estimation. Still some researchers proposed the introduction of a new kind of fuzzy subsets to remedy the problem. The range of these new fuzzy subsets is not the unit interval but the following set

$$I([0, 1]) = \left\{(a, b) \mid (a, b \in [0, 1]) \wedge (a \leq b)\right\}.$$

Thus, when are asked to what degree Amelia is tall, we reply by giving two numbers that are the two ends of an interval and so our estimation is a range of numbers instead of a single number. These *new* fuzzy subsets are called *interval valued* fuzzy sets have been proposed by Roland Sambuc [51]:

Definition 3.11.1 An interval valued fuzzy set A is a function $A : X \to I([0, 1])$, where X is some universe, and $A(x) = [A_*(x), A^*(x)]$. In particular, the values returned by the two functions A_* and A^* are the lower and upper end of the membership degree and for all $x \in X$: $A_*(x) \leq A^*(x)$.

Obviously, if $A_*(x) = A^*(x)$ for all $x \in X$, then A is an ordinary fuzzy set. To put it differently, ordinary fuzzy sets are a special cases of interval valued fuzzy sets. The basic operations between these sets are presented below.

Definition 3.11.2 Assume that $A, B : X \rightarrow I([0, 1])$ are two interval valued fuzzy subsets. Then,

- their *union* is the interval valued fuzzy subset with the following property

$$(A \cup B)(x) = \left[\max\left(A_*(x), B_*(x)\right), \max\left(A^*(x), B^*(x)\right) \right];$$

- their *intersection* is the interval valued fuzzy subset with the following property

$$(A \cap B)(x) = \left[\min\left(A_*(x), B_*(x)\right), \min\left(A^*(x), B^*(x)\right) \right];$$

- the *complement* of A is the fuzzy interval valued set

$$A^{\complement}(x) = \left[1 - A^*(x), 1 - A_*(x) \right].$$

3.12   Fuzzy Multisets

The idea that an element belongs to a subset to a certain degree is not the only way to extend sets and their theory. By violating the rule that any element belongs 0 or 1 time to a set and replacing it with another one according to which an element may belong to a set 0, 1, or more times, we get a new kind of set. This new mathematical structure is known in the literature as *multiset*. These structures are not mathematical curiosities but correspond to real things. For example, consider a quadratic equation $ax^2 + bx + c = 0$, where a, b, and c are real numbers. This equation has always two solutions and we know that if the discriminant (i.e., the quantity $b^2 - ac$) is equal to zero, then the equation has two real solution that are equal. Another example where multisets are useful is the prime factorization of integers. Recall, that a prime number is an integer above 1 that cannot be made by multiplying other integer numbers. For example, 17 is a prime number. Given an integer number x, prime factorization is finding which prime numbers multiply together to make x. For instance, we can easily verify that $90 = 2 \times 3^2 \times 5$. So, the prime factors of 90 are 3, 3, 2 and 5, that is, a multiset. A more practical example of a multiset is a shopping cart. Suppose we are visiting a supermarket and we put out shopping cart 3 boxes of McVitie's family circle biscuits, 2 packs of Maxima Green Recycled Toilet Tissue, etc. Clearly, the contents of the shopping cart form a multiset. Formally, we define multisets as follows:

Definition 3.12.1 Let X be a *universe* (i.e., an arbitrary set). A *multiset A* of X, is realized by a function $A : X \rightarrow \mathbb{N}$, which is called its *membership function*. For every $x \in X$, the value $A(x)$ denotes the *number of times x belongs to the multiset A*.

If you compare the previous definition with the definition of fuzzy subsets (see Definition 3.2.1), then you will notice that the two definitions are similar. This is quite reasonable since, in both cases, we change the notion membership.

Let us consider again the shopping cart. This time we put products that belong to the supermarkets' own brand labels and "ordinary" products. A number of people believe that own brand label products are inferior products when compared to "ordinary" products, nevertheless, whether this claim is true or not is not something I will discuss here. If we put in our cart 4 boxes of own brand family circle biscuits and 3 boxes McVitie's family circle biscuits, then we will have in our shopping cart 3 boxes of biscuits that are family circle biscuits with degree equal to 1.0 and 4 boxes of biscuits that are family circle biscuits to degree equal to, say, 0.9. Our shopping cart can be described mathematically with a structure that is called fuzzy multiset.[6] Yager [74] defined fuzzy multisets as follows.

Definition 3.12.2 Assume X is a set of elements. Then a fuzzy bag [i.e, a multiset] A drawn from X can be characterized by a function Count.Mem$_A$ such that

$$\text{Count.Mem}_A : X \rightarrow Q,$$

where Q is the set of all crisp bags drawn from the unit interval.

From this definition we conclude that a fuzzy multiset is a function from a a set X and each element of x is mapped to a multiset of $[0, 1]$. In different words, a fuzzy multiset is a function that maps elements of X to elements of $\mathbb{N}^{[0,1]}$.

In the simplest case, *currying* transforms a function that takes two arguments to a function that takes one argument and returns a function that takes one argument. In particular, if we have a function

$$f : X \times Y \rightarrow Z,$$

then we can transform it to thew following function

$$\hat{f} : X \rightarrow Z^Y.$$

---

[6] The gentle reader may consider the previous example artificial in the sense that we have two different products. However, there are many products out there that look identical, feel identical, but are a bit different (e.g., consider the two famous brands of cola drinks). Thus, if we have plate of biscuits that are from two different boxes that look identical, it is quite possible that our guests may not be able to say which biscuit comes from which box, although some may be able to spot the difference in taste.

Currying is named after Haskell Brooks Curry [15] but it was developed by Moses Ilyich Schönfinkel [49] (Моисей Эльевич Шейнфинкель):

> Some contemporary logicians call this way of looking at a function "currying," because I made extensive use of it; but Schönfinkel had the idea some 6 years before I did.

However, it seems that the idea was introduced by Friedrich Ludwig Gottlob Frege. The opposite operation is called *uncurrying* and transforms function $\hat{f}$ to function f. Since a fuzzy multiset A is a function

$$A : X \rightarrow \mathbb{N}^{[0,1]},$$

we can use uncurrying, to get the following function

$$A : X \times [0, 1] \rightarrow \mathbb{N}.$$

However, it is more natural to demand that for each element x there is only one membership degree and one multiplicity. In other words, a "fuzzy multiset" A should be described by a function $X \rightarrow [0, 1] \times \mathbb{N}$. To distinguish these structures from fuzzy multisets, I call them *multi-fuzzy* sets [60, 61].

Suppose that A is multi-fuzzy set. Then, we can define the following two functions: the *multiplicity* function $A_m : X \rightarrow \mathbb{N}$ and the *membership* function $A_\mu : X \rightarrow [0, 1]$. If $A(x) = (n, i)$, then $A_m(x) = n$ and $A_\mu(x) = i$.

Remark 3.12.1 Any ordinary set $A \subseteq X$ is identical to the multi-fuzzy set $\bar{A}$ defined as follows:

$$\bar{A}(a) = (1, \chi_A(a)), \text{ for all } a \in X,$$

In addition, any fuzzy set $A : X \rightarrow [0, 1]$ is identical to the multi-fuzzy set A' defined as follows:

$$A'(a) = (A(a), 1), \text{ for all } a \in X,$$

Similarly, any multiset $M : X \rightarrow \mathbb{N}$ can be represented by the multi-fuzzy set:

$$\mathscr{M}(a) = (1, M(a)), \text{ for all } a \in X.$$

Let me now define the union, the intersection, the sum and the difference of two multi-fuzzy sets.

Definition 3.12.3 Assume that A and B are two multi-fuzzy sets with universe the set Z, then their union is:

$$(A \cup B)(z) = \left(\max\left\{ A_m(z), B_m(z) \right\}, \max\left\{ A_\mu(z), B_\mu(z) \right\} \right).$$

Notice that in the case of multisets the union is defined in terms of the max operator. Also, the typical definition of fuzzy subset intersection is given in terms of max. Thus, the definition above is fully justified. Similarly, since the operations of set intersection for both multisets and fuzzy subsets is defined in terms of the min operator, the following definition is completely reasonable:

Definition 3.12.4 Assume that A and B are two multi-fuzzy sets with universe the set Z, then their intersection is:

$$(A \cap B)(z) = \left(\min\{A_m(z), B_m(z)\}, \min\{A_\mu(z), B_\mu(z)\} \right).$$

The sum is an operation that is defined only for multisets and it is actually a generalization of the union, it really makes sense to use the max operator to define the membership degrees, while the multiplicity is defined as usual.

Definition 3.12.5 Assume that A and B are two multi-fuzzy sets with universe the set Z, then their sum is:

$$(A \uplus B)(z) = \left(A_m(z) + B_m(z), \max\{A_\mu(z), B_\mu(z)\} \right).$$

For reasons similar to the previous case, the difference of two multi-fuzzy sets is defined as follows:

Definition 3.12.6 Assume that A and B are two multi-fuzzy sets with universe the set Z, then their difference is:

$$(A \ominus B)(z) = \left(\max\{A_m(z) - B_m(z), 0\}, \min\{A_\mu(z), B_\mu(z)\} \right).$$

3.13 The Extension Principle

In Sect. 2.2, I introduced functions between ordinary sets. Fuzzy subsets are mathematical structures that extend or *generalize* sets. Therefore, it is quite natural to ask: Can we define functions and other related structures between fuzzy subsets? The answer to this question is affirmative and in this section I will explain how we can define functions between fuzzy subsets (the exposition that follows is based on the corresponding expositions found in [39] and [63]).

Let us consider the function $f : \mathbb{R} \to \mathbb{R}$ that doubles its argument (i.e., $f(x) = 2x$). Also, consider the fuzzy subset "about 2" that is depicted in Fig. 3.3. If there is a fuzzy

Fig. 3.3 The visual representation of a fuzzy subset "about 2"

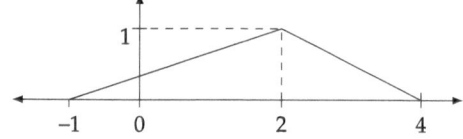

function that doubles its argument, what is the result of applying this function to the fuzzy subset "about 2"?

First of all, it is important to realize that this new function will not map a single value to a single value (e.g., the number 2 to the number 4) but the it will map the function $V : \mathbb{R} \rightarrow [0, 1]$, which is the fuzzy subset "about 2," to another function, which is another fuzzy subset. Technically, we want to find the image f when it is applied to D. But first let us see what is the image of an ordinary function. Given a function $g : A \rightarrow B$, its *image* is a set that contains all elements $b \in B$ that are assigned by g:

$$\mathrm{Im}(g) = \big\{ b \mid a \in A, \ g(a) = b, \ \text{and } b \in B \big\}.$$

Although this definition looks quite clear, many students and even some teachers of mathematics fail to see the difference between the codomain and the image of a function. Now suppose that $h : \mathbb{R} \rightarrow \mathbb{R}$ is a function. Then, according to the previous definition every single element $a \in \mathbb{R}$ is associated with a unique element $b \in \mathbb{R}$. The element b is called the image of a under h. The image of h is the set that contains all such images. Suppose that $h(a) = 4$, then each a is associated with 4, thus, the image of h is the set $\{4\}$. In general, when $h(a) = b$, we can find the image of h by solving the equation $b = h(a)$ and consider a as the unknown quantity. Then, all values of b that solve this equation form the image of h. For example, if $h(a) = 3/a$, then we have to solve the equation $b = 3/a$. In this case, $a = 3/b$, which means the image of h is the set $\mathbb{R} \setminus \{0\}$, since a is not defined when $b = 0$.

Suppose that $M \subset A$, then what is the image $g[M]$ of M under g? In this case, it is the set that contains all elements $b \in B$ such that $g(a) = b$, for all elements of $a \in M$. In different "words":

$$b \in g[M] \text{ if and only if there is an } a \in A \text{ such that } a \in M \text{ and } g(a) = b. \qquad (3.2)$$

It is not difficult to see that $\mathrm{Im}(M) \subseteq \mathrm{Im}(g)$. For example, if $M = [0, 2]$, $g : \mathbb{R} \rightarrow \mathbb{R}$, and $g(x) = x^2$, then $g[M] = [0, 4]$.

The next logical step is to use "equation" 3.2 to define the image of a fuzzy subset $D : A \rightarrow [0, 1]$ under a function $g : A \rightarrow B$. To do so, we transform 3.2 as follows

$$b \in g[D] \text{ if and only if there is an } a \in A \text{ such that } D(a) \text{ \textbf{and} } g(a) = b. \qquad (3.3)$$

In the previous "equation" I use the word **and** to emphasize that here we have a logical *conjunction*. But what is conjunction? Let me briefly explain this logical operation. When we have two sentences that are connected with the word *and*, then the result is a sentence whose meaning depends on the meaning of the two sentences. In particular, if both of them are true, then the whole sentence is true. For example, if I have the clauses "Today is Wednesday" and "Today's weather is sunny" and both are true, then the sentence "Today is Wednesday and the weather is sunny" is also true. However, if one the two clauses is false, then the whole sentence is false. Naturally, if both clauses are false, the while sentence is false. If the sentence "John entered the building" is false and the sentence "Mary left the building" is true, then the sentence "John entered the building and Mary left it" is false. In computers and in fuzzy mathematics, we denote true by the number 1 and false by the number 0. Thus, in "equation" 3.3 we actually have a fuzzy conjunction since $D(a)$ is number that belongs to the unit interval and $g(a) = b$ is either true or false. Function min plays the role of the conjunction operator in fuzzy logic and we can use it to transform "equation" 3.3 to the following one.

$$\llbracket b \in g[D] \rrbracket \text{ if and only if there is an } a \in A \text{ such that } \min\big(D(a), \llbracket g(a) = b \rrbracket\big), \quad (3.4)$$

where $\llbracket v \rrbracket$ denotes the truth value of the sentences v. In particular, if v is true, then $\llbracket v \rrbracket = 1$; otherwise $\llbracket v \rrbracket = 0$. The "if and only if" part of the previous equation can be replaced by the symbol "=":

$$\llbracket b \in g[D] \rrbracket = \text{there is an } a \in A \text{ such that } \min\big(D(a), \llbracket g(a) = b \rrbracket\big). \quad (3.5)$$

When we say that "there is an $a \in A$ such that $P(a)$" and $A = \{a_1, a_2, \ldots, a_n\}$, we actually mean that the statement

$$P(a_1) \textbf{ or } P(a_2) \textbf{ or } \cdots \textbf{ or } P(a_n)$$

is true. In this formula **or** is the *disjunction* operator. When we have two sentences P and Q and we want to find whether the sentence "P **or** Q" is true, we just need to examine whether P or Q is true. If one of them is true, then the whole sentence is also true. More generally, a sentence like the previous one is true, if one of its sub-sentences (i.e., $P(a_k)$ in this case) is true. In fuzzy logic, function max plays the role of the disjunction operator. Thus, the formula above becomes $\max\big(P(a_1), P(a_2), \ldots, P(a_n)\big)$ when $P(a_k)$ are fuzzy sentences. Therefore, "equation" 3.5 should be transformed to the following one:

$$g[D](b) = \max\big\{\min\big(D(a_1), \llbracket g(a_1) = b \rrbracket\big), \ldots, \min\big(D(a_n), \llbracket g(a_n) = b \rrbracket\big)\big\}. \quad (3.6)$$

Since all bs are in general real numbers, I have to say that there are cases where we cannot find the maximum of a set of real numbers. Indeed, this sounds strange but think of the set of all negative real numbers. This set does not have a maximum as there is no negative number y with the property that $x \le y$ for all negative numbers x. To fully understand this, think of the sequence of negative numbers $-0.1, -0.01, -0.001$, etc. Although there is no negative number that is greater of all negative numbers, there are many numbers that are greater than all of them. For example, the numbers 0, 0.001, 0.01, 0.1 are all greater than any negative number. Since there are many such numbers, the smallest number that is greater than all negative numbers is called the *supremum* of the set of negative numbers. More generally, for any set S its supremum is a number t such that $s \leq t$ for all $s \in S$. Note that the maximum of a set must belong to the set while the supremum of the set is not necessarily a member of it. Also, the dual notion of the supremum is called *infimum*. We can use these ideas and facts to transform "equation" 3.6 to the following one.

$$g[D](b) = \sup\{(D(a) \mid g(a_n) = b\}.\tag{3.7}$$

This equation can be used to solve our original problem and it is known as the *extension principle*. In simple words, the extension principle says that the membership degree $b \in B$ to the image of the fuzzy set D under the mapping $g : A \to B$ is the greatest possible $D(a)$ for all a that are mapped to b under g. If we are sure that there is only one a_k such that $[\![g(a_k) = b]\!]$ is true, and we want to find the image of a a fuzzy subset whose universe is a finite set, then we can use the following definition,

Definition 3.13.1 Suppose that

$$D = \left\{ \frac{d_1}{a_1} + \frac{d_2}{a_2} + \cdots + \frac{d_n}{a_n} \right\},$$

is a fuzzy subset of A and $g : A \to B$ is a function. Then, the image of D under g is the following fuzzy subset of B:

$$g[D] = g\left(\left\{ \frac{d_1}{a_1} + \frac{d_2}{a_2} + \cdots + \frac{d_n}{a_n} \right\} \right)$$

which is equivalent to

$$g[D] = \left\{ \frac{d_1}{g(a_1)} + \frac{d_2}{g(a_2)} + \cdots + \frac{d_n}{g(a_n)} \right\}.$$

Example 3.9 Suppose that $A = \{1, 2, 3, 4, 5\}$ and $B = \{1, 2, \ldots, 199, 200\}$ and $g : A \to B$ is a function that returns the cube of its argument, that is, $g(a) = a^3$. Also, assume that the following is a fuzzy subset of A:

$$\text{small} = \left\{ \frac{1.0}{1} + \frac{0.95}{2} + \frac{0.8}{3} + \frac{0.6}{4} + \frac{0.4}{5} \right\}.$$

Then,

$$g[\text{small}] = \text{small}^3 = \left\{ \frac{1.0}{1} + \frac{0.95}{8} + \frac{0.8}{27} + \frac{0.6}{48} + \frac{0.4}{125} \right\}.$$

■

Example 3.10 Suppose we have the following fuzzy subset

$$A = \left\{ \frac{0.1}{-2} + \frac{0.4}{-1} + \frac{0.8}{0} + \frac{0.9}{1} + \frac{0.3}{2} \right\}$$

and the function $f(x) = x^2 - 3$. Then, $f[A]$ is the following fuzzy subset:

$$f[A] = \left\{ \frac{0.8}{-3} + \frac{\max(0.4, 0.9)}{-2} + \frac{\max(0.1, 0, 3)}{1} \right\}$$

$$= \left\{ \frac{0.8}{-3} + \frac{0.9}{-2} + \frac{0.3}{1} \right\}.$$

∎

Let us see how we can find the image of a fuzzy subset whose universe is not a finite set. We will work with the fuzzy subset "about 2," which is depicted in Fig. 3.3, and the function $f(a) = 2a$. It should be clear that this function does not map two different elements to the same value. However, this is not true for $h(a) = a^2$. To understand how we do find the image of the fuzzy subset under f, I will show how we find the membership degrees $f["about\ 2"](b)$ for $b \in \{-2, 4, 8\}$.

$$f["about\ 2"](-2) = "about\ 2"(-1) = 0$$
$$f["about\ 2"](4) = "about\ 2"(2) = 1$$
$$f["about\ 2"](8) = "about\ 2"(4) = 0$$

If we use function h instead of f, then let see how we compute a single value:

$$h["about\ 2"](1) = \max\{"about\ 2"(-1), "about\ 2"(1)\} = 0.85,$$

provided that "about 2"$(1) = 0.85$. Figure 3.4 shows the result of this operation.

Suppose that we have two fuzzy subsets $A : X \to [0, 1]$ and $B : X \to [0, 1]$ and we have a function $h : X \times X \to X$. Then, how can we find $h[A, B]$? Since we view ordinary operations like addition or multiplication as functions that take two arguments, solving this problem will allow us to solve problems like the addition of two fuzzy subsets of the set $\mathbb{R}$.

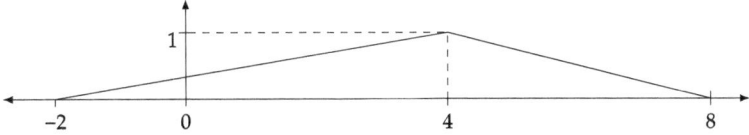

Fig. 3.4 The visual representation of a fuzzy subset $h["about\ 2"]$

The problem stated in the previous paragraph can be solved by extending the notion of the Cartesian product to fuzzy subsets. Suppose we have the fuzzy subsets $A_1 : X_1 \to [0, 1]$, $A_2 : X_2 \to [0, 1],\ldots, A_n : X_n \to [0, 1]$. Then, their Cartesian product is the subset

$$A_1 \times A_2 \times \cdots A_n, \tag{3.8}$$

where

$$(A_1 \times A_2 \times \cdots A_n)(x_1, x_2, \ldots, x_n) = \min\{A_1(x_1), A_2(x_2), \ldots, A_n(x_n)\}.$$

This definition allows us to *extend* the extension principle to functions with two, three, or more arguments. Here is how we can solve the problem stated above. Consider the function

$$g : X_1 \times X_2 \times \cdots \times X_n \to Y.$$

Then, the image of the Cartesian product (3.8) under g is the fuzzy subset

$$g[A_1, A_2, \ldots, A_n] = g[A_1 \times A_2 \times \cdots A_n].$$

Thus, if we want to compute $g[A_1, A_2, \ldots, A_n](y)$, then we have the extended version of the extension principle that follows:

$$g[A_1, A_2, \ldots, A_n](y) =$$
$$\sup_{(x_1,\ldots,x_n)\in X_1\times\cdots\times X_n} \{(A_1 \times A_2 \times \cdots A_n)(x_1 \ldots, x_n) \mid g(x_1, \ldots, x_n) = y\} =$$
$$\sup_{(x_1,\ldots,x_n)\in X_1\times\cdots\times X_n} \{\min(A_1(x_1), \ldots, A_n(x_n)) \mid g(x_1, \ldots, x_n) = y\}.$$
$$\tag{3.9}$$

Example 3.11 Suppose that we have the following fuzzy subsets:

$$A = \left\{\frac{0.3}{4} + \frac{1.0}{6} + \frac{0.2}{8}\right\} \quad \text{and} \quad B = \left\{\frac{0.1}{12} + \frac{0.6}{14} + \frac{0.8}{18}\right\},$$

and the function $f(a, b) = 3a - b + 5$. Then, we can easily compute the corresponding values of function f:

$f(4, 12) = 5$	$f(4, 14) = 3$	$f(4, 18) = -1$
$f(6, 12) = 11$	$f(6, 14) = 9$	$f(6, 18) = 5$
$f(8, 12) = 17$	$f(8, 14) = 15$	$f(8, 18) = 11$

The following table shows how we calculate the initial membership degrees.

$A \diagdown B$	$\dfrac{0.1}{12}$	$\dfrac{0.6}{14}$	$\dfrac{0.8}{18}$
$\dfrac{0.3}{4}$	$\dfrac{0.1}{5}$	$\dfrac{0.1}{3}$	$\dfrac{0.1}{-1}$
$\dfrac{1.0}{6}$	$\dfrac{0.1}{11}$	$\dfrac{0.6}{9}$	$\dfrac{0.8}{5}$
$\dfrac{0.2}{8}$	$\dfrac{0.1}{17}$	$\dfrac{0.2}{15}$	$\dfrac{0.2}{11}$

Using these data, we can find the fuzzy subset $f[A, B]$:

$$
\begin{aligned}
f[A, B] &= \left\{ \frac{0.1}{-1} + \frac{0.1}{3} + \frac{\max(0.1, 0.8)}{5} + \frac{0.6}{9} + \frac{\max(0.1, 0.2)}{11} + \frac{0.1}{12} + \frac{0.6}{14} + \frac{0.2}{15} + \frac{0.1}{17} + \frac{0.8}{18} \right\} \\
&= \left\{ \frac{0.1}{-1} + \frac{0.1}{3} + \frac{0.8}{5} + \frac{0.6}{9} + \frac{0.2}{11} + \frac{0.1}{12} + \frac{0.6}{14} + \frac{0.2}{15} + \frac{0.1}{17} + \frac{0.8}{18} \right\}
\end{aligned}
$$

∎

Fuzzy Numbers and Linguistic Variables

4

> *As for numbers, they hate nobody and nobody can afford to hate them.*
>
> *—Shakuntala Devi*
> *(Indian mental calculator and writer)*

Simply put, a fuzzy number is any fuzzy subset of the set of real numbers. Fuzzy numbers are quite useful in the mathematical formulation of vague statements about real numbers. Thus, the fuzzy subset "about 2" that was discussed in Sect. 3.13 is a fuzzy number. There are fuzzy systems where some of their *variables* range over states that are fuzzy numbers. Typically, these fuzzy numbers may represent terms such as "quite big," "very big," "somewhat small," etc. If this is the case, then these variables are called linguistic variables.

4.1   What Is a Fuzzy Number?

Although I have given a rough description of fuzzy numbers, still it is necessary to give a precise definition.

Definition 4.1.1 A fuzzy subset of $\mathbb{R}$ is a fuzzy number provided it has the following properties:

(1) they are normal fuzzy subsets;
(2) their supports are bounded (i.e., they are not infinite); and
(3) their α-cuts are closed intervals for every $a \in (0, 1]$.

Recall that a fuzzy subset whose maximal membership value is equal to one is called normal. Therefore, a fuzzy subset of $\mathbb{R}$ which is not normal is not a fuzzy number. Of course, this is

© The Author(s), under exclusive license to Springer Nature Switzerland AG 2025
A. Syropoulos, *Fuzzy Mathematics*, Synthesis Lectures on Mathematics & Statistics,
https://doi.org/10.1007/978-3-031-73834-0_4

more than obvious but my point is that that there many fuzzy subsets of $\mathbb{R}$ that are not fuzzy numbers. There are positive and negative fuzzy numbers:

Definition 4.1.2 A fuzzy number A is called positive (negative), if $A(x) = 0$, for all $x < 0$ $(x > 0)$.

In a way, all real numbers are fuzzy numbers and here is why. Take a real number a and let us say we can identify it with the singleton $\{a\}$. It is known that every subset $S \subset \mathbb{R}$ is uniquely determined by giving its characteristic function. In the case of $\{a\}$ it holds that

$$\chi_a(x) = 1 \text{ iff } x = a \text{ and } \chi_a(x) = 0 \text{ iff } x \neq a.$$

Thus, the characteristic function of the singleton defines a fuzzy number that corresponds to a real number.

4.2   Shapes of Fuzzy Numbers

From all fuzzy numbers those that have specific shapes are particularly interesting. In what follows, I will describe the fuzzy numbers with shapes that have found most applications in science and engineering.

4.2.1   Triangular Fuzzy Numbers

Among the various shapes of fuzzy numbers, triangular fuzzy numbers are the most common ones. But let us see why they are called triangular. Suppose we are given a fuzzy number and we are told that is triangular. Then, we need to take a sheet of graph paper (i.e., a sheet of paper that has a specific layout of squares) to draw its graph. In general, the *graph* of a function is the collection of all ordered pairs of the function. that is, if $f : A \rightarrow B$ is a function, then all pairs (a, b) such that $f(a) = b$ are its graph. Clearly, in most cases we cannot compute the graph of a function, so we construct a table of a number of pairs that we use to draw the graph. In our case, we construct such a table for the fuzzy number. Obviously, the more pairs we have, the better the final graph. Next, we plot the axes of a Cartesian coordinate system where our data can fit and we draw dots whose coordinates are taken from the table. Finally, we connect the dots and the result is the graph of the fuzzy number. In our case, the result will be a triangle.

Usually, we specify a triangular fuzzy number $A : \mathbb{R} \rightarrow [0, 1]$ using the notation tfn$(\overline{x}, l, r)$ [25], where $A(\overline{x}) = 1$, l is the length of the line segment from x to the vertex that lies to the left of it, and r is the length of the line segment from x to the vertex that lies to the right of it. Figure 4.1 shows the graph of tfn$(3, 3, 2)$. The membership function of a triple $(\overline{x}, l, r)$ that defines a triangular fuzzy number is given below:

Fig. 4.1 A triangular fuzzy number

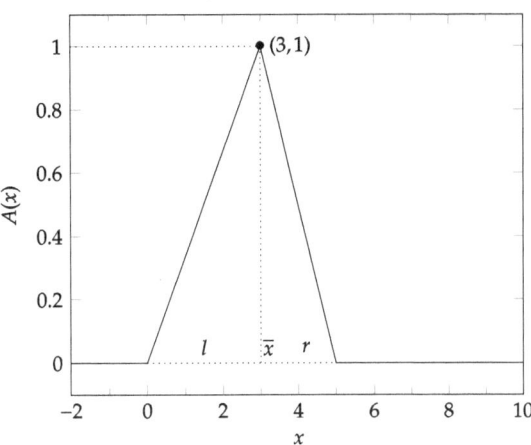

Alternatively, this membership function can be expressed as follows [25]:

$$A(x) = \min\Big[\max\big[0,\, 1 - (\overline{x} - x)/l\big],\, \max\big[0,\, 1 - (x - \overline{x})/r\big]\Big], \quad \text{for all } x \in \mathbb{R}. \qquad (4.1)$$

In fact, this alternative formulation has been used to draw the various figures in this section.

In Sect. 4.3 I will describe a general method that can be used to compute the sum, the difference, the product, or the quotient of two or more fuzzy numbers of any shape, but before that, I will say a few things about the arithmetic operations of triangular fuzzy numbers. Given two such numbers $A = \text{tfn}(\overline{x_1}, l_1, r_1)$ and $B = \text{tfn}(\overline{x_2}, l_2, r_2)$, we know that:

(i) The sum and the difference of A and B are also a triangular fuzzy numbers.
(ii) The symmetric image of A is also a triangular fuzzy number.
(iii) The product and the quotient of A and B are not triangular fuzzy numbers. but we can approximate them with triangular fuzzy numbers [58].
(iv) The maximum and the minimum of A and B are not triangular fuzzy numbers.
(v) $A + B = \text{tfn}(\overline{x_1} + \overline{x_2}, l_1 + l_2, r_1 + r_2)$.
(vi) $A - B = \text{tfn}(\overline{x_1} - \overline{x_2}, |l_1 - l_2|, |r_1 - r_2|)$, where $|x| = x$ when $x > 0$ and $|x| = -x$ when $x < 0$.
(vii) The symmetric image of A is $-A = \text{tfn}(-\overline{x_1}, l_1, r_1)$.

(viii) According to [58], the following is an approximation of the product of two triangular fuzzy numbers:

$$A \times B \approx \text{tfn}\Big(\overline{x_1} \cdot \overline{x_2}, \min(l_1 l_2, l_1 r_2, r_1 l_2, r_1 r_2), \max(l_1 l_2, l_1 r_2, r_1 l_2, r_1 r_2)\Big).$$

(ix) Again, according to [58], the following is an approximation of the quotient of two triangular fuzzy numbers:

$$A/B = \text{tfn}\Big(\overline{x_1}/\overline{x_2}, \min(l_1/l_2, l_1/r_2, r_1/l_2, r_1/r_2), \max(l_1/l_2, l_1/r_2, r_1/l_2, r_1/r_2)\Big).$$

Note that [45] presents exact formulas for the computation of the product and the quotient of triangular fuzzy numbers, nevertheless, I have applied their method for two simple simple fuzzy numbers and it failed. I have not checked the other solutions that are presented in [45], but I guess they are not correct either. Figure 4.2 shows the sum $\text{tfn}(3, 3, 2) + \text{tfn}(1, 4, 3)$ and Fig. 4.3 shows the difference $\text{tfn}(3, 3, 2) - \text{tfn}(1, 4, 3)$. In addition, using the formulas above I have verified that

$$\text{tfn}(3, 3, 2) - \text{tfn}(1, 4, 3) = \text{tfn}(3, 3, 2) \times \text{tfn}(1, 4, 3).$$

Fig. 4.2 Graphical representation of $\text{tfn}(3, 3, 2) + \text{tfn}(1, 4, 3)$

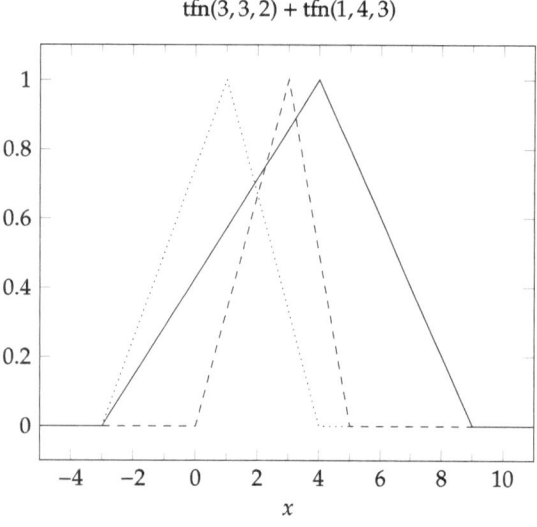

Fig. 4.3 Graphical representation of $\mathrm{tfn}(3, 3, 2) - \mathrm{tfn}(1, 4, 3)$

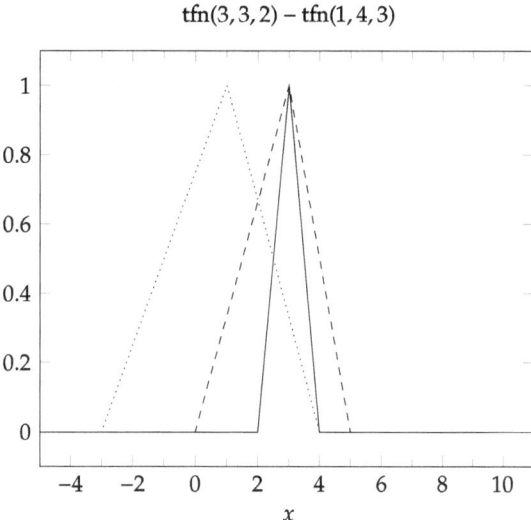

$$\mathrm{tfn}(3, 3, 2) - \mathrm{tfn}(1, 4, 3)$$

4.2.2 Trapezoidal Fuzzy Numbers

A *trapezoidal* fuzzy number A is a fuzzy subset of $\mathbb{R}$ for which there is more than one element $x \in \mathbb{R}$ such that $A(x) = 1$. Strictly speaking, these fuzzy subsets are not fuzzy numbers but they are interval valued fuzzy subsets. A trapezoidal fuzzy set is completely characterized by four real numbers $t_1 \le t_2 \le t_3 \le t_4$. I will use the notation $\mathrm{trfn}(t_1, t_2, t_3, t_4)$ to specify the fuzzy subset that is characterized by the following membership function:

$$T(x) = \begin{cases} 0, & \text{if } x < a \\ (x - a)/(b - a), & \text{if } a \le x < b \\ 1, & \text{if } b \le x < c \\ (d - x)/(d - c), & \text{if } c \le x < d \\ 0, & \text{if } x \ge d \end{cases}$$

Figure 4.4 shows the trapezoidal fuzzy subset $(15, 25, 55, 90)$. Note that the four numbers of the quadruple correspond to the x-coordinates of the four vertices of the resulting "trapezoid" in this specific order.

In [69] the authors describe the operations between trapezoidal fuzzy subsets. They assume that all the numbers that identify the fuzzy subsets are positive. Here is how they define addition and multiplication:

(i) $\mathrm{trfn}(t_1, t_2, t_3, t_4) + \mathrm{trfn}(s_1, s_2, s_3, s_4) = \mathrm{trfn}(t_1 + s_1, t_2 + s_2, t_3 + s_3, t_4 + s_4)$.

Fig. 4.4 A trapezoidal fuzzy
number

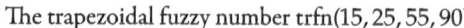

The trapezoidal fuzzy number trfn(15, 25, 55, 90)

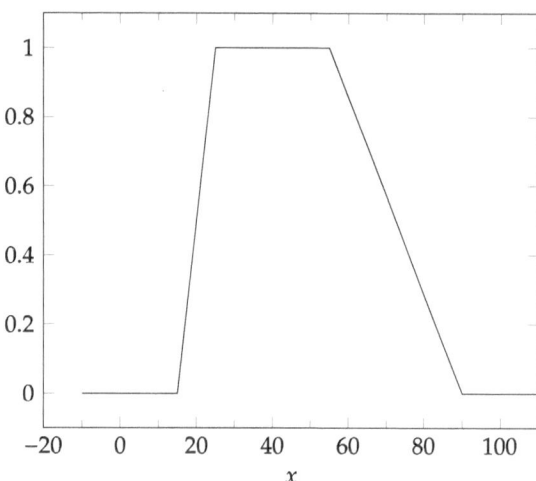

(ii) $\mathrm{trfn}(t_1, t_2, t_3, t_4) \times \mathrm{trfn}(s_1, s_2, s_3, s_4) = \mathrm{trfn}(r_1, r_2, r_3, r_4)$, where $T_1 = \{t_1 \times s_1, t_1 \times s_2, t_2 \times s_1, t_2 \times s_2\}$, $T_2 = \{t_3 \times s_2, t_3 \times s_4, t_4 \times s_3, t_4 \times s_4\}$, and $r_1 = \min(T_1)$, $r_2 = \max(T_1)$, $r_3 = \min(T_2)$, and $r_4 = \max(T_2)$.

They also define the square root or the absolute value of a fuzzy subset, but these operations are quite involved. The interested reader is very welcome to consult [69]. Figure 4.5 shows two trapezoidal fuzzy numbers and their sum.

Fig. 4.5 The sum of two
trapezoidal fuzzy numbers

trfn(15, 25, 55, 90) + trfn(10, 30, 60, 110)

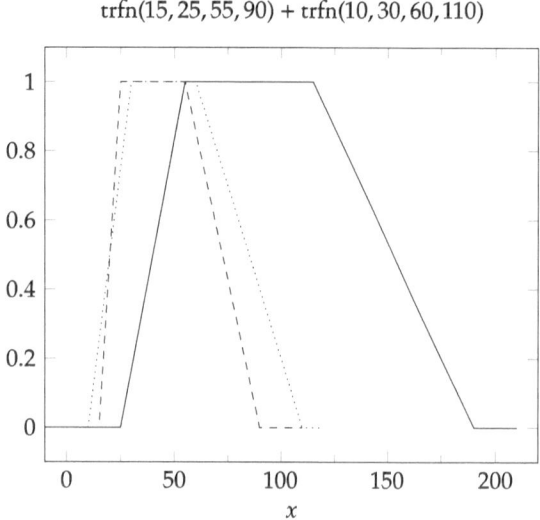

4.2.3 Gaussian Fuzzy Numbers

The membership function of a *Gaussian* fuzzy number looks like some special kind of a Gaussian function. In general, the Gaussian function is the *probability density* function of the *normal distribution*.[1] The following figure shows the base form of this function.

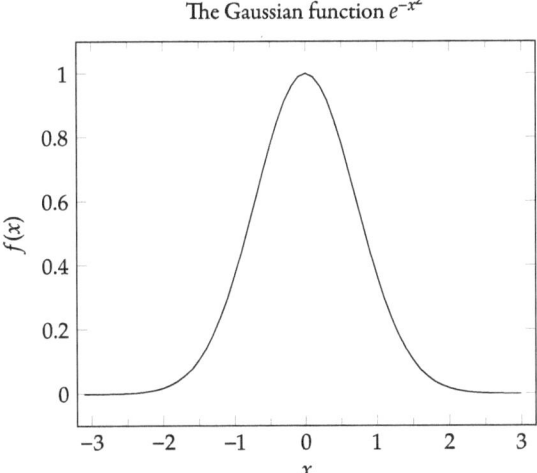

The Gaussian function e^{-x^2}

The general form of the membership function of a Gaussian fuzzy number is:

$$f(x) = e^{-\frac{(x-b)^2}{2c^2}},$$

where $e \approx 2.718281828459045$ is Napier's constant which is also known as Euler's number. When the exponent of e is a lengthy expression, we write $\exp(X)$ instead of e^X. In general, one specifies a Gaussian fuzzy number using the notation $\mathrm{gfn}(\overline{x}, \sigma_l, \sigma_r)$, where $\overline{x}$ is the only element whose membership degree is equal to 1, and σ_l and σ_r correspond to the left-hand and right-hand distance from the point that has membership degree equal to 1, see Fig. 4.6. The membership function of any Gaussian fuzzy number has the following general form:

$$G(x) = \begin{cases} \exp[-(x-\overline{x})^2/(2\sigma_l^2)], & \text{if } x < \overline{x} \\ \exp[-(x-\overline{x})^2/(2\sigma_r^2)], & \text{if } x \geq \overline{x} \end{cases} \tag{4.2}$$

A *quasi-Gaussian* fuzzy number consists of a Gaussian fuzzy number whose membership degree is set to zero for $x < \overline{x} - 3\sigma_l$ and for $x > \overline{x} + 3\sigma_r$, respectively. A quasi-Gaussian fuzzy number will be written as $\mathrm{gfn}^*(\overline{x}, \sigma_l, \sigma_r)$ and, in general, the following function is the general form of the membership function any quasi-Gaussian fuzzy number:

---

[1] I will not try to explain what these terms mean. After all, it is not my intension to provide a thorough introduction to probability theory.

Fig. 4.6 A Gaussian fuzzy
number

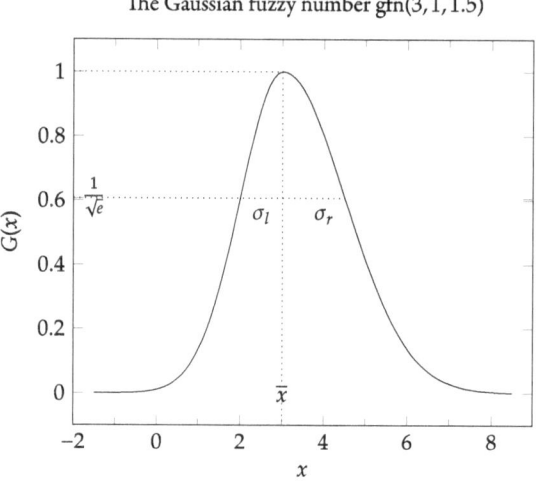

The Gaussian fuzzy number gfn(3, 1, 1.5)

$$G_q(x) = \begin{cases} 0, & \text{if } x \leq \overline{x} - 3\sigma_l \\ \exp\left[-(x-\overline{x})^2/(2\sigma_l^2)\right], & \text{if } \overline{x} - 3\sigma_l < x < \overline{x} \\ \exp\left[-(x-\overline{x})^2/(2\sigma_r^2)\right], & \text{if } \overline{x} \leq x < \overline{x} + 3\sigma_r \\ 0, & \text{if } x \geq \overline{x} + 3\sigma_r \end{cases}$$

Note that the fuzzy number has non-zero membership values within a specific range.

4.2.4   Quadratic Fuzzy Numbers

A *quadratic* fuzzy number is one more shape of a fuzzy number. These fuzzy numbers are using the notation qfn($\overline{x}, \beta_l, \beta_r$). The membership function of a quadratic fuzzy number depends on three parameters:

$$Q(x) = \begin{cases} 0, & \text{if } x \leq \overline{x} - \beta_l \\ 1 - (x-\overline{x})^2/\beta_l^2 & \text{if } x - \beta_l < x < \overline{x} \\ 1 - (x-\overline{x})^2/\beta_r^2 & \text{if } \overline{x} \leq x < \overline{x} + \beta_r \\ 0, & \text{if } x \geq \overline{x} - \beta_r \end{cases}$$

Figure 4.7 shows exactly to what these three parameters correspond.

Fig. 4.7 A quadratic fuzzy
number

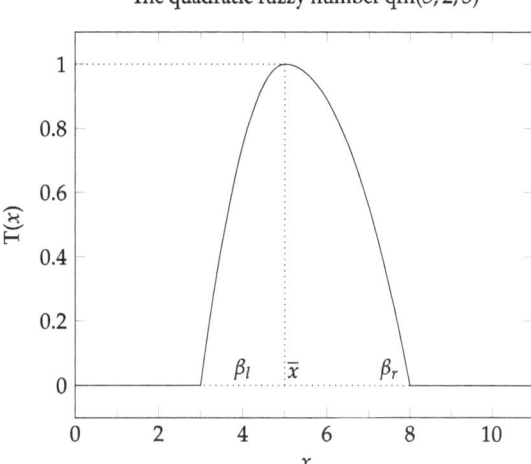

The quadratic fuzzy number qfn(5, 2, 3)

4.2.5 Exponential Fuzzy Numbers

An *exponential* fuzzy number is the last shape of fuzzy numbers that I will briefly present.
Their membership function has the general form:

$$
E(x) = \begin{cases}
0, & \text{if; } x < \overline{x} - a\tau_l \\
\exp\left[(x - \overline{x})/\tau_l\right], & \text{if } \overline{x} - a\tau_l \leq x < \overline{x} \\
\exp\left[(\overline{x} - x)/\tau_r\right], & \text{if } \overline{x} \leq x < \overline{x} + a\tau_r \\
0, & \text{if; } \overline{x} + a\tau_r \leq x
\end{cases}
$$

This definition is borrowed from [6] and here τ_l and τ_r are the left and right spread of $\overline{x}$
respectively, and a represents a tolerance value. I will specify an exponential fuzzy number
using the notation efn$(\overline{x}, \tau_l, \tau_r, a)$. Figure 4.8 shows an exponential fuzzy number.

4.2.6 LR Fuzzy Numbers

The last shape of fuzzy numbers that I am going to briefly present are the so-called L-R
fuzzy numbers. The graph of the membership function of these fuzzy numbers consists of
two parts—a left curve and right curve that meet at point $(\overline{x}, 1)$. Figure 4.9 shows a typical
example of an L-R fuzzy number. Below, I give the most general form of the membership
function of an L-R fuzzy number.

Fig. 4.8 An exponential fuzzy number

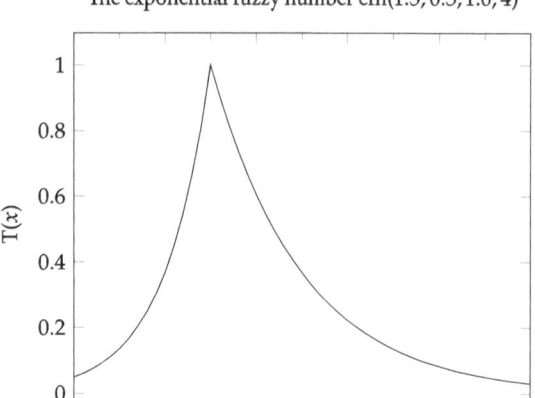

The exponential fuzzy number efn(1.5, 0.5, 1.0, 4)

Fig. 4.9 An L-R fuzzy number

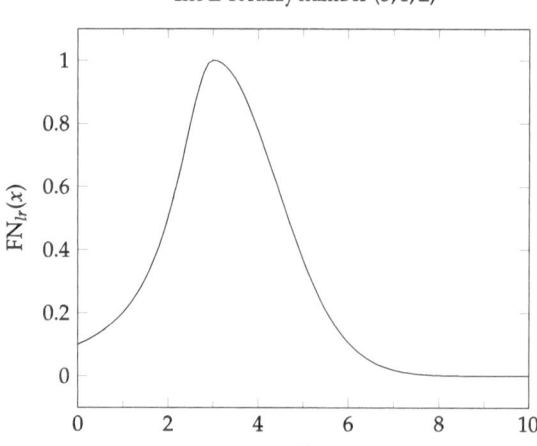

The L-R fuzzy number ⟨3, 1, 2⟩

$$A(x) = \begin{cases} L\left(\frac{\overline{x}-x}{\alpha}\right), & \text{if } x \leq \overline{x} \text{ and } \alpha > 0 \\[2mm] R\left(\frac{x-\overline{x}}{\beta}\right), & \text{if } x \geq \overline{x} \text{ and } \beta > 0. \end{cases}$$

Note that L and R are two functions that must have the following properties for all $0 \leq x \in \mathbb{R}$:

1. $L(x) \in [0, 1]$ and $R(x) \in [0, 1]$ for all x;
2. $L(0) = R(0) = 1$;
3. $L(-x) = L(x)$ and $R(-x) = R(x)$; and

4. L and R are *decreasing* in $[0, +\infty)$, that is, for amy two number x and y in $[0, +\infty)$ such that $x < y$, we have $L(x) \geq L(y)$ and $R(x) \geq R(y)$.

For example, in the case of the fuzzy number shown in Fig. 4.9 I have used:

$$\bar{x} = 3$$
$$\alpha = 1$$
$$\beta = 2$$
$$L(x) = \frac{1}{1 + x^2}$$
$$R(x) = e^{-x^2}$$

4.2.7   Generalized Fuzzy Numbers

A *generalized* fuzzy number $A = \text{gfn}(a, b, c, d; w)$ is a fuzzy subset of $\mathbb{R}$ that satisfies the following conditions:

1. A is a continuous function, that is, a function whose graph does not have any breaks or jumps;
2. $A(x) = 0$ when $-\infty < x \leq a$, $-\infty$ denotes minus infinity, that is, a "number" representing a quantity with no bounds that is less than every real number;
3. A is *strictly increasing* in $[a, b]$, that is, for any two number $x, y \in [a, b]$ such that $x < y$, it holds that $A(x) < A(y)$;
4. $A = w$ when $b \leq x \leq c$;
5. A is strictly decreasing in $[c, d]$, that is, for any two number $x, y \in [c, d]$ such that $x > y$, it holds that $A(x) > A(y)$; and
6. $A(x) = 0$ when $d \leq x < \infty$, where ∞ is the opposite of $-\infty$.

Here w is called the *degree of confidence* of some expert's opinion. If $w = 1$, then the generalized fuzzy number A is a normal trapezoidal fuzzy number. If $a = b$ and $c = d$, then A is a crisp interval. If $b = c$, then A is a generalized triangular fuzzy number. If $a = b = c = d$ and $w = 1$, then A is a real number.

4.3   Arithmetic of Fuzzy Numbers

I have already presented some methods by which one can compute the sum or the difference of fuzzy numbers that have a specific shape. Naturally, it does make sense to be able to compute the sum, the difference, the product, and the quotient of any two or more fuzzy numbers, but the question is how can we compute these fuzzy numbers? Suppose that $\circledast$ denotes any of the four arithmetic operators and A, B two arbitrary fuzzy numbers, then there are two methods that can be used to compute $A \circledast B$: The first one makes use of arithmetic operations between intervals and the second one is using the extension principle.

4.3.1 Interval Arithmetic

Let us start by explaining the arithmetic operations between intervals. The four arithmetic operations on closed intervals are defined as follows:

$$[a, b] + [d, e] = [a + d, b + e]$$
$$[a, b] - [d, e] = [a - e, b - d]$$
$$[a, b] \times [d, e] = \big[\min(a \times d, a \times e, b \times d, b \times e), \max(a \times d, a \times e, b \times d, b \times e)\big]$$

and, provided that $0 \notin [d, e]$,

$$[a, b] \div [d, e] = [a, b] \times [1/e, 1/d]$$
$$= \big[\min(a \div d, a \div e, b \div d, b \div e), \max(a \div d, a \div e, b \div d, b \div e)\big].$$

Any real number r can be considered to be the degenerated interval $[r, r]$. For example, if a, b, and s are positive numbers, then

$$[a, b] \times [s, s] = \big[\min(a \times s, a \times s, b \times s, b \times s), \max(a \times s, a \times s, b \times s, b \times s)\big]$$
$$= \big[\min(a \times s, b \times s), \max(a \times s, b \times s)\big]$$
$$= \big[a \times s, b \times s\big].$$

In addition, the following equalities can be easily proved.

$$[a, b] + [0, 0] = [0, 0] + [a, b] = [a, b] \text{ and } [a, b] \times [1, 1] = [1, 1] \times [a, b] = [a, b].$$

Example 4.1 Consider the intervals $[2, 5]$ and $[4, 7]$, then

$$[2, 5] + [4, 7] = [2 + 4, 5 + 7] = [6, 12]$$
$$[2, 5] - [4, 7] = [2 - 7, 5 - 4] = [-5, 1]$$
$$[2, 5] \times [4, 7] = \big[\min(2 \times 4, 2 \times 7, 5 \times 4, 5 \times 7), \max(2 \times 4, 2 \times 7, 5 \times 4, 5 \times 7)\big]$$
$$= \big[\min(8, 14, 20, 35), \max(8, 14, 20, 35)\big] = [8, 35]$$
$$[2, 5] \div [4, 7] = \big[\min(2 \div 4, 2 \div 7, 5 \div 4, 5 \div 7), \max(2 \div 4, 2 \div 7, 5 \div 4, 5 \div 7)\big]$$
$$= \Big[\min\Big(\frac{1}{2}, \frac{2}{7}, \frac{5}{4}, \frac{5}{7}\Big), \max\Big(\frac{1}{2}, \frac{2}{7}, \frac{5}{4}, \frac{5}{7}\Big)\Big] = [0.5, 1.25]$$

4.3.2   Interval Arithmetic and α-Cuts

We know that all α-cuts of a fuzzy number are intervals that are closed and both their endpoints are real numbers. In addition, we know from Theorem 3.3.1 that the union of all α-cuts of any fuzzy set make up the set itself. Thus, knowing all α-cuts of an arithmetic operation between two fuzzy numbers means that we can compute the result of the computation.

Consider two fuzzy numbers A and B and suppose that $\circledast$ is one of the four arithmetic operations. Then, the fuzzy subset $A \circledast B$ is defined using its α-cut $^{\alpha}(A \circledast B)$ as

$$^{\alpha}(A \circledast B) = {}^{\alpha}A \circledast {}^{\alpha}B$$

for all $\alpha \in (0, 1]$. In case we want to divide two fuzzy numbers, it is necessary to ensure that $0 \notin {}^{\alpha}A$ for all $\alpha \in (0, 1]$. In general, it would be useful to be able to use all α-cuts. For instance, if $^{\alpha}A = [a_1(\alpha), a_2(\alpha)]$ and $^{\alpha}B = [b_1(\alpha), b_2(\alpha)]$, then

$$^{\alpha}(A + B) = [a_1(\alpha) + b_1(\alpha), a_2(\alpha) + b_2(\alpha)].$$

So far it has been explained how one can use α-cuts to compute, say, the sum of two fuzzy numbers, but, nothing has been said about the details of the method. Instead of giving a general recipe, I will give a concrete example that demonstrates the general method (the ideas presented in the example were first published in [33]).

Example 4.2 Suppose we have two Gaussian fuzzy numbers $A = \text{gfn}(4, 0.8, 1.0)$ and $B = \text{gfn}(3, 1.5, 2.0)$. If we want to compute an α-cut of, say, A, we need to compute the following interval:
$$^{\alpha}A = [a_1(\alpha), a_2(\alpha)],$$

where

$$a = \exp\left[-(a_1(\alpha) - \overline{x})^2/(2\sigma_l^2)\right] \quad \text{and} \quad b = \exp\left[-(a_2(\alpha) - \overline{x})^2/(2\sigma_r^2)\right].$$

This is justified by the fact that the left end of the interval should be described by the first case of the function Definition 4.2 because $x < \overline{x}$ and the right end will be described by the second case of the same function definition because $x \geq \overline{x}$. Now let us compute $a_1(\alpha)$:

$$a = \exp\left[-(a_1(\alpha) - \overline{x})^2/(2\sigma_l^2)\right] \Longleftrightarrow$$

$$\ln(a) = \frac{-(a_1(\alpha) - \overline{x})^2}{2\sigma_l^2} \Longleftrightarrow$$

$$2\ln(a) = \frac{-(a_1(\alpha) - \overline{x})^2}{\sigma_l^2} \Longleftrightarrow$$

$$-2\ln(a) = \frac{(a_1(\alpha) - \overline{x})^2}{\sigma_l^2} \Longleftrightarrow$$

$$\sqrt{-2\ln(a)} = a_1(\alpha) - \overline{x} \Longleftrightarrow$$

$$\sigma_l\sqrt{-2\ln(a)} = a_1(\alpha) - \overline{x} \Longleftrightarrow$$

$$\overline{x} + \sigma_l\sqrt{-2\ln(a)} = a_1(\alpha)$$

The symbol $\Longleftrightarrow$ means that the equalities are equivalent. From the last equation we get that

$$a_1(\alpha) = \overline{x} + \sigma_l\sqrt{-2\ln(a)}$$

Similarly, we have:

$$b = \exp\left[-(a_2(\alpha) - \overline{x})^2/(2\sigma_r^2)\right].$$

After some calculations we get

$$a_2(\alpha) = \overline{x} + \sigma_4\sqrt{-2\ln(b)}$$

Thus, the α-cut of $^\alpha A$ is the following interval

$$\left[\overline{x} + \sigma_l\sqrt{-2\ln(a)}, \overline{x} + \sigma_4\sqrt{-2\ln(b)}\right]$$

The sum of two such intervals is

$$\left[(\overline{x'} + \overline{x''}) + (\sigma_l' + \sigma_l'')\sqrt{-2\ln(a)}, (\overline{x'} + \overline{x''}) + (\sigma_r' + \sigma_r'')\sqrt{-2\ln(b)}\right]$$

Now we need to go back to the membership function. In this case, we solve the following equations for a and b:

$$x = (\overline{x'} + \overline{x''}) + (\sigma_l' + \sigma_l'')\sqrt{-2\ln(a)}$$

$$x = (\overline{x'} + \overline{x''}) + (\sigma_r' + \sigma_r'')\sqrt{-2\ln(b)}$$

Let us solve the first equation:

$$x = (\overline{x'} + \overline{x''}) + (\sigma_l' + \sigma_l'')\sqrt{-2\ln(a)} \Longleftrightarrow$$

$$x - (\overline{x'} + \overline{x''}) = (\sigma_l' + \sigma_l'')\sqrt{-2\ln(a)} \Longleftrightarrow$$

$$\frac{x - (\overline{x'} + \overline{x''})}{(\sigma_l' + \sigma_l'')} = \sqrt{-2\ln(a)} \Longleftrightarrow$$

$$\frac{(x - (\overline{x'} + \overline{x''}))^2}{(\sigma_l' + \sigma_l'')^2} = -2\ln(a) \Longleftrightarrow$$

$$\frac{-(x - (\overline{x'} + \overline{x''}))^2}{2(\sigma_l' + \sigma_l'')^2} = \ln(a) \Longleftrightarrow$$

$$a = \exp\left[\frac{-\left(x - (\overline{x'} + \overline{x''})\right)^2}{2(\sigma'_l + \sigma''_l)^2}\right].$$

Solving the second equation gives

$$b = \exp\left[\frac{-\left(x - (\overline{x'} + \overline{x''})\right)^2}{2(\sigma'_r + \sigma''_r)^2}\right].$$

Clearly, the values of a and b correspond to the values of the two cases of the membership function of the sum of two Gaussian fuzzy numbers. I have use these equations to draw the fuzzy numbers gfn(4, 0.8, 1.0), gfn(3, 1.5, 2.0), and gfn(4, 0.8, 1.0) + gfn(3, 1.5, 2.0) in Fig. 4.10. ∎

4.3.3 Fuzzy Arithmetic and the Extension Principle

Arithmetic operations between fuzzy numbers can also be defined using the extension principle. Roughly, we "transform" operations between real numbers into operations between fuzzy real numbers. Suppose that A and B are two fuzzy numbers. Then, we can define the four arithmetic operations between A and B plus the min and max operations for all $z \in \mathbb{R}$ as follows:

Fig. 4.10 The sum of two Gaussian fuzzy numbers—the dashed line is the number gfn(4, 0.8, 1.0), the dotted line is the number gfn(3, 1.5, 2.0), and the continuous line is their sum

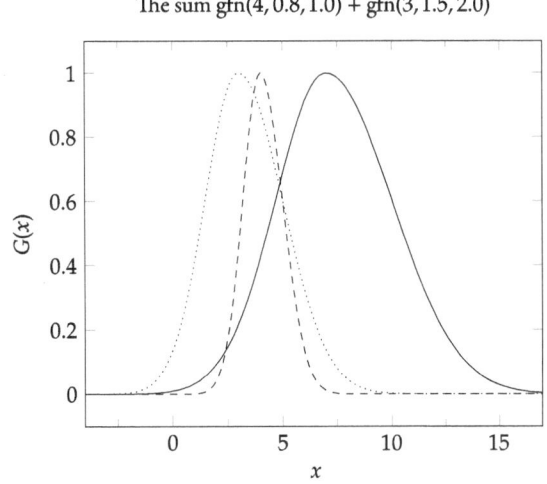

The sum gfn(4, 0.8, 1.0) + gfn(3, 1.5, 2.0)

$$(A + B)(z) = \sup_{z=x+y} \min[A(x), B(y)],$$

$$(A - B)(z) = \sup_{z=x-y} \min[A(x), B(y)],$$

$$(A \cdot B)(z) = \sup_{z=x \cdot y} \min[A(x), B(y)],$$

$$(A/B)(z) = \sup_{z=x/y} \min[A(x), B(y)],$$

$$(\min(A, B))(z) = \sup_{z=\min(x,y)} \min[A(x), B(y)],$$

$$(\max(A, B))(z) = \sup_{z=\max(x,y)} \min[A(x), B(y)].$$

The two methods that can be used to perform arithmetic operations between fuzzy numbers are equivalent. However, this method is best suited for discrete fuzzy numbers [71]:

Definition 4.3.1 A fuzzy set $A : \mathbb{R} \to [0, 1]$ is a *discrete fuzzy number* if A has *finite* support $x_1 < x_2 < \cdots < x_n$ and there are indices $s, t, 1 \leq s \leq t \leq n$ such that

(i) $A(x_i) = 1$ whenever $s \leq i \leq t$,
(ii) if $i < j < s$, then $A(x_i) \leq A(x_j) < 1$,
(iii) if $t > p > q$, then $1 > A(x_p) \geq A(x_q)$.

An explanation of this definition is in order. The first condition means that there is at least on element x_i for which $A(x_i) = 1$. Also, for all elements that are to the left of x_i, the membership degrees of A are "sorted," that is, $A(x_i) \leq A(x_j)$ when x_i is to the left of x_j, and also all membership degrees are less than one. Finally, for all elements that are to the right of x_i, the membership degrees of A are "sorted" inversely, that is, $A(x_i) \geq A(x_j)$ when x_i is to the right of x_j, and also all membership degrees are, again, less than one. For example, the following is a discrete fuzzy number

$$A = \frac{0.3}{1} + \frac{0.4}{2} + \frac{1.0}{3} + \frac{0.8}{4} + \frac{0.7}{5},$$

while the following is not a discrete fuzzy number

$$B = \frac{0.4}{5} + \frac{0.2}{7} + \frac{1.0}{9} + \frac{0.5}{11} + \frac{0.6}{13}.$$

Example 4.3 Consider the two discrete fuzzy numbers $A = 0.3/1 + 1.0/3 + 0.7/5$ and $B = 0.2/7 + 1.0/9 + 0.5/11$. Let us compute $\max(A, B)$. First, we examine the various values that z takes.

- We see that $7 = \max(1, 7)$, $7 = \max(3, 7)$, and $7 = \max(5, 7)$. Therefore, we first compute the minimums:

$$\min[A(1), B(7)] = \min[0.3, 0.2] = 0.2$$
$$\min[A(3), B(7)] = \min[1.0, 0.2] = 0.2$$
$$\min[A(5), B(7)] = \min[0.7, 0.2] = 0.2$$

And now we compute the membership degree of 7:

$$\big(\max(A, B)\big)(7) = \max(0.2, 0.2, 0.2) = 0.2.$$

- We see that $9 = \max(1, 9)$, $9 = \max(3, 9)$, and $9 = \max(5, 9)$. We compute the minimums:

$$\min[A(1), B(9)] = \min[0.3, 1, 0] = 0.3$$
$$\min[A(3), B(9)] = \min[1.0, 1.0] = 1.0$$
$$\min[A(5), B(9)] = \min[0.7, 1.0] = 0.7$$

Let's compute the membership degree of 9:

$$\big(\max(A, B)\big)(9) = \max(0.3, 1.0, 0.7) = 1.0.$$

- We see that $11 = \max(1, 11)$, $11 = \max(3, 11)$, and $11 = \max(5, 11)$. Let's find the minimums:

$$\min[A(1), B(11)] = \min[0.3, 0.5] = 0.3$$
$$\min[A(3), B(11)] = \min[1.0, 0.5] = 0.5$$
$$\min[A(5), B(11)] = \min[0.7, 0.5] = 0.5$$

And now we need to find the membership degree of 11:

$$\big(\max(A, B)\big)(11) = \max(0.3, 0.5, 0.5) = 0.5.$$

Therefore, $\max(A, B) = 0.2/7 + 1.0/9 + 0.5/11$. ∎

Note that sometimes the operations between discrete fuzzy numbers yield fuzzy sets that are not discrete fuzzy numbers. Also, the technique just described can be extended to non-discrete fuzzy numbers but it is more difficult to proceed.

Example 4.4 Consider again the discrete fuzzy numbers $A = 0.3/1 + 1.0/3 + 0.7/5$ and $B = 0.2/7 + 1.0/9 + 0.5/11$. This time, let us compute $A + B$. We examine the various values that z takes.

- We see that $8 = 1 + 7$. Therefore, $\min[A(1), B(7)] = \min[0.3, 0.2] = 0.2$ and the membership degree of 8 is
$$(\max(A, B))(8) = 0.2.$$

- We see that $10 = 1 + 9$ and $10 = 3 + 7$. Thus,
$$\min[A(1), B(9)] = \min[0.3, 1.0] = 0.3$$
$$\min[A(3), B(7)] = \min[1.0, 0.2] = 0.2$$

Therefore,
$$(\max(A, B))(10) = \max(0.3, 0.2) = 0.3.$$

- We see that $12 = 1 + 11$, $12 = 3 + 9$, and $10 = 5 + 7$. Thus,
$$\min[A(1), B(11)] = \min[0.3, 0.5] = 0.3$$
$$\min[A(3), B(9)] = \min[1.0, 1.0] = 1.0$$
$$\min[A(5), B(7)] = \min[0.7, 0.2] = 0.2$$

Therefore,
$$(\max(A, B))(12) = \max(0.3, 1.0, 0.2) = 1.0.$$

- We see that $14 = 3 + 11$ and $14 = 5 + 9$. Thus,
$$\min[A(3), B(11)] = \min[1.0, 0.5] = 0.5$$
$$\min[A(5), B(9)] = \min[0.7, 1.0] = 0.7$$

Therefore,
$$(\max(A, B))(14) = \max(0.5, 0.7) = 0.7.$$

- We see that $16 = 5 + 11$. Therefore, $\min[A(5), B(11)] = \min[0.7, 0.5] = 0.5$ and the membership degree of 16 is
$$(\max(A, B))(16) = 0.5.$$

Therefore, $A + B = 0.2/8 + 0.3/10 + 1.0/12 + 0.7/14 + 0.5/16$. Verify that this is indeed a discrete fuzzy number! ■

4.4 Linguistic Variables

Apparently, the term "linguistic variable" has multiple meanings. For example, in linguistics a *linguistic variable* [40] is defined to be a set of related dialect forms (i.e., words, idioms, etc.) that mean the same thing and which correlate with some social grouping in a speech community [11]. However, the term "linguistic variable" has a specific meaning in fuzzy mathematics and it was introduced by Zadeh in a series of papers [79–81].

We know that in mathematics a variable is a name that stands for a value from a collection (not necessary a set) of different values. We have seen variables that stand for elements of sets, for sets, for numbers, etc., but we have not seen variables to stand for words, usually adjectives, or sentences. Consider the attribute temperature (e.g., of a room). Then, the words freezing, chilly, frigid, cold, cool, warm, hot, very hot, scorching, sweltering, blistering, roasting, and boiling describe the relative warmth or coolness of the room or the environment, in general. For each of these *linguistic variables* we can define a membership function for a specific temperature range. Since these ranges may overlap, the same temperature is mapped to multiple membership values in the range of 0 to 1 by each function. These membership values can then be used to say whether a temperature is hot, cold, or cool. Figure 4.11 depicts some of these membership functions. Here the membership functions are trapezoidal fuzzy numbers. In general, one can use any kind of fuzzy number but this kind of numbers seemed ideal for this case.

Formally, a linguistic variable is characterized by a quintuple

$$(V, T(V), U, G, M),$$

Fig. 4.11 Representation of linguistic variables of temperature

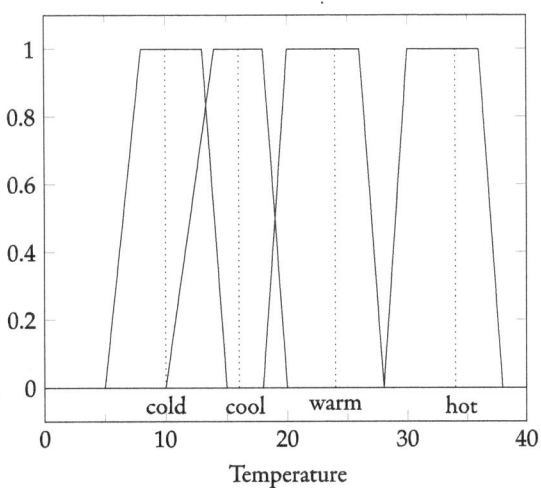

where

- V is the name of the variable;
- $T(V)$ is the set of terms of V, that is, a set of linguistic values of V, which are fuzzy sets on the universe X;
- U is a universe;
- G is a syntactic rules for generating the names of values of V; and
- M is a semantic rules for associating each value with its meaning, that is, the membership function that characterizes the fuzzy set.

Since this definition is not so easy to gulp down at once, I will explain the meaning of each part of the quintuple. Let us continue with temperatures. Then, V is TEMPERATURE and $T(V)$ can be a set containing the adjectives or phrases given in the previous paragraph, or any other term derived from them, or any other term that can be used to describe a temperature. Consider the linguistic term "warm" of the linguistic variable temperature that is defined as

$$M = \left\{ \left(x, \text{TEMPERATURE}_{\text{warm}}(x)\right) \,\middle|\, x \in [0, 55] \right\}$$

and

$$\text{TEMPERATURE}_{\text{warm}}(x) = \text{trfn}(18, 20, 26, 28).$$

The syntactical rules G that generate the modified value of the linguistic terms of x is actually what we call a *formal language*. Such a language consists of a set of *terminal* symbols (e.g., letters, words, etc.) that is called an *alphabet*, a set of *non-terminal* symbols (i.e., a symbol that can be broken down to a sequence that consists of terminal and/or other non-terminal symbols), a starting non-terminal symbol, and a set of *production rules*, that are used to break down a starting symbol to a sequence of terminal symbols. In fact, the production rules are used to examine if a specific sequence of symbols can be derived from the starting symbol using the specific production rules. Let me give an example, which I hope will make things clearer. Suppose we are given the sentence "Peter plays golf" and we are asked to check whether this sentence is a valid English language sentence. The English language can be viewed as a formal language and the starting non-terminal symbol is S, which stands for sentence. The set of terminal symbols include all the words of the English language, punctuation symbols, etc. The set of non-terminal symbols include symbols like V or NP, which stand for verb and noun phrase, respectively. Figure 4.12 shows how we can analyze the sentence and check that indeed "Peter plays golf" is a valid English language sentence. More specifically, we start from S, which is broken down to a NP and a VP (verb phrase, this is actually a production rule). Then, NP is substituted by the terminal symbol "Peter." The symbol V is broken down to a V and a NP. These symbols are substituted by the words "plays" and "golf," respectively. Since now we have only terminal symbols, we can say that "Peter plays golf" is a valid sentence.

Fig. 4.12 The syntactic tree of
a simple sentence

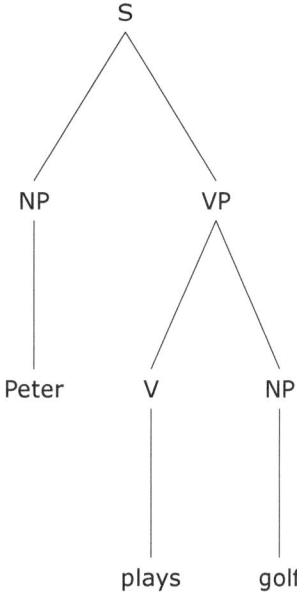

4.5   Linguistic Hedges

Linguistic hedge have been define by George Lakoff [41] as follows[2]:

> For me, some of the most interesting questions are raised by the study of words whose meaning
> implicitly involves fuzziness—words whose job is to make things fuzzier or less fuzzy. I will
> refer to such words as "hedges."

Some words or phrases that are hedges are: *sort of*, *kind of*, *loosely speaking*, *more or less*,
roughly, *very*, *often*, *mostly*, etc. Since hedges are used in oral and written communication,
Zadeh [78] gave a mathematical treatment of hedges as functions that modify linguistic
variables.

Linguistic hedges are primarily used to express in a more precise form the degree of
correctness and truth of a particular statement. For example, consider the statement "Lucy is
young" and assume that is has truth degree equal to 0.7, then the truth degree of the statement
"Lucy is very young" is equal to $0.7^2 = 0.49$. In addition, the statement "Lucy is not very
young" has a truth degree that is equal to $1 - 0.49 - 0.51$. If X is statement and $T(X)$ is its
truth degree, then the word very is modeled by the function $very(x) = x^2$. However, hedges
are more interesting when we apply them to fuzzy *predicates*. A predicate is a statement
whose truth value belongs to [0, 1] and involves an argument. For example, tall(x) is a

---

[2] This paper was originally published in *The Chicago Which Hunt: Papers from the Relative Clause
Festival*, pp. 183–228, 1972.

predicate and its truth value depends on the value of x. Thus, when $x = 1.75$ m, tall$(1.75$ m$)$ is the truth degree to which a person having height 1.75 m is tall. In other words, a predicate (fuzzy or crisp) is statement of the form: x is P. Consider the fuzzy predicate $F(x)$ and a linguistic hedge H, then the modified proposition $H(F(x))$ is the result of applying H to $F(x)$. Since, the truth value $F(A)$, where A is an concrete object, number, person, etc., belongs to $[0, 1]$, the hedge must be realized by a function $h : [0, 1] \to [0, 1]$. For example, the hedges *very* and *fairly* are realized by the functions

$$h_{\text{very}}(x) = x^2 \quad \text{and} \quad h_{\text{fairly}}(x) = \sqrt{x}.$$

These functions and all other functions that are used in the same way are called *modifiers*.

Modifiers are either *strong* or *weak*. A modifier h is strong when $h(a) < a$ and weak when $h(a) > a$. Consider again the statement tall$(1.75$ m$)$ whose truth value is 0.65. Then, the statement "a person having height 1.75 m is very tall" has truth degree equal to $0.65 \times 0.65 = 0.4225$. This means that h_{very} is strong. On the other hand, the statement "a person having height 1.75 m is fairly tall" has truth degree equal to $\sqrt{0.65} = 0.806225774829855$, which implies that h_{fairly} is weak.

A function h that has the following characteristics can be used as a modifier:

- h is a continuous function, that is, when its graph is a single unbroken curve.
- $h(0) = 0$ and $h(1) = 1$.
- If h is strong, then $1/h$ is weak and vice versa.
- If h and g are two modifiers then *compositions* (i.e., the functions $h(g(x))$ and $g(h(x))$) of these modifiers are also modifiers. For example, if h stands "very" and g stands for "often," then $h \circ g$ stands for "very often." Note that $(h \circ g)(x) = f(g(x))$.

A class of functions that satisfy these conditions is the class

$$h_\alpha(x) = x^\alpha,$$

where α is a positive real number. When $\alpha < 1$, then h_α is a weak modifier and when $\alpha > 1$, then h_α is a strong modifier. The modifier h_1 is the identity modifier, that is the modifier that does nothing.

Bibliography

1. Khurshid Ahmad and Andrea Mesiarova-Zemankova. "Choosing t-Norms and t-Conorms for Fuzzy Controllers". In: *Fourth International Conference on Fuzzy Systems and Knowledge Discovery (FSKD 2007)*. Vol. 2. 2007, pp. 641–646.
2. Kassimrir T. Atanassov. "Intuitionistic Fuzzy Sets". In: *Fuzzy Sets and Systems* 20.1 (1986), pp. 87–96.
3. Krassimir T. Atanassov. *Intuitionistic Fuzzy Sets: Theory and Applications*. Heidelberg: Physica-Verlag HD, 1999.
4. Krassimir T. Atanassov. *On Intuitionistic Fuzzy Sets Theory*. Berlin: Springer Berlin Heidelberg, 2012.
5. Nicholas of Autrecourt. *The Universal Treatise*. Translated by Leonard A. Kennedy, Richard E. Arnold, and Arthur E. Millward. Milwaukee, Wisconsin: The Marquette University Press, 1971.
6. Barnabás Bede. *Mathematics of Fuzzy Sets and Fuzzy Logic*. Studies in Fuzziness and Soft Computing 295. Berlin: Springer-Verlag, 2013.
7. Gleb Beliakov. "9 - Fitting triangular norms to empirical data". In: *Logical, Algebraic, Analytic and Probabilistic Aspects of Triangular Norms*. Ed. by Erich Peter Klement and Radko Mesiar. Amsterdam: Elsevier Science B.V., 2005, pp. 261–272.
8. James C. Bezdek. "Fuzzy models—What are they, and why? [Editorial]". In: *IEEE Transactions on Fuzzy Systems* 1.1 (1993), pp. 1–6.
9. David Bloor. *Knowledge and Social Imagery*. 2nd. Chicago 60637: The University of Chicago Press, 1991.
10. Nicolas Bourbaki. *Theory of Sets*. Originally published as "Éléments de Mathématique: Théorie des Ensembles" by Hermann, Publishers in Arts and Science, Paris, 1968. Berlin: Springer, 2013.
11. David Britain. *Sociolinguistic variation*. The LLAS Centre for Languages, Linguistics and Area Studies. 2016. URL: https://web-archive.southampton.ac.uk/www.llas.ac.uk/resources/gpg/1054.html#toc_2 (visited on 06/21/2024).
12. Otávio Bueno and Mark Colyvan. "Just what is Vagueness?" In: *Ratio: An international journal of analytic philosophy* 25 (2012), pp. 19–33.
13. Florian Cajori. *A history of mathematical notations*. "This is an unabridged and unaltered republication in one volume of the work first published in two volumes by The Open Court Publishing Company, La Salle, Illinois, in 1928 and 1929." Mineola, N.Y: Dover Publications, 1993.

A. Syropoulos, *Fuzzy Mathematics*, Synthesis Lectures on Mathematics & Statistics, https://doi.org/10.1007/978-3-031-73834-0

14. B. Jack Copeland. "On Vague Objects, Fuzzy Logic and Fractal Boundaries". In: *The Southern Journal of Philosophy* 33.S1 (1995), pp. 83–96.

15. Haskell Brooks Curry. "Some Philosophical Aspects of Combinatory Logic". In: *The Kleene Symposium*. Ed. by Jon Barwise, H. Jerome Keisler, and Kenneth Kunen. Vol. 101. Studies in Logic and the Foundations of Mathematics. Elsevier, 1980, pp. 85–101.

16. Sujit Das et al. "Neutrosophic fuzzy set and its application in decision making". In: *Journal of Ambient Intelligence and Humanized Computing* 11.11 (2020), pp. 5017–5029.

17. Didier Dubois. "Have fuzzy sets anything to do with vagueness?" In: *Understanding Vagueness: Logical, Philosophical, and Linguistic Perspectives*. Ed. by Petr Cintula et al. London: College Publications, 2012, pp. 317–346.

18. Didier Dubois et al. "Terminological difficulties in fuzzy set theory—The case of 'Intuitionistic Fuzzy Sets' ". In: *Fuzzy Sets and Systems* 156.3 (2005). 40th Anniversary of Fuzzy Sets, pp. 485–491.

19. Euclid. *Euclid's Elements*. Translated by Thomas Little Heath and edited by Dana Densmore. Santa Fe, New Mexico: Green Lion Press, 2007.

20. Gareth Evans. "Can there be vague objects?" In: *Analysis* 38.4 (1978), p. 208.

21. Robin Giles. "The concept of grade of membership". In: *Fuzzy Sets and Systems* 25.3 (1988). Interpretation of Grades of Membership, pp. 297–323.

22. Siegfried Gottwald. "Fuzzy Set Theory: Some Aspects of the Early Development". In: *Aspects of Vagueness*. Ed. by Heinz J. Skala, S. Termini, and E. Trillas. Dordrecht, The Netherlands: D. Reidel Publishing Company, 1984, pp. 13–29.

23. Paul Richard Halmos. *Naive Set Theory*. New York: Springer Science+Business Media, 1974.

24. Peter Hanks. "Bipolarity and Sense in the Tractatus". In: *Journal for the History of Analytical Philosophy* 2.9 (2014), pp. 1–15.

25. Michael Hanss. *Applied Fuzzy Arithmetic: An Introduction with Engineering Applications*. Berlin: Springer, 2005.

26. Felix Hausdorff. *Set Theory*. 2nd. Translation from the original German, into English, of the 3rd (1937) ediion of *Mengenlehre* (translator: John R. Aumann, et al.). New York: Chelsea Publishing Company, 1962.

27. Thomas Little Heath. *A History of Greek Mathematics*. Volume I: From Thales to Euclid. Oxford, UK: Oxford University Press, 1921.

28. Zuzana Heinzová and Ján Belko. "Scrutinizing Pupil's Social Status by Means of a Sociometric Questionnaire". In: *The New Educational Review* 5.1 (2005), pp. 45–50.

29. Arend Heyting. *Intuitionism: An Introduction*. 3rd. Amsterdam: North-Holland Pub. Co., 1976.

30. Randall Holmes. *Elementary Set Theory with a Universal Set*. Vol. 10. Cahiers du Centre de logique, Academia-Bruylant. Louvain-la-Neuve, Belgium: Editions ACADEMIA-EME, 1998.

31. Dominic Hyde and Mark Colyvan. "Paraconsistent Vagueness: Why Not?" In: *Australasian Journal of Logic* 6 (2008), pp. 107–121.

32. International Organisation for Standardization. *ISO 5725-1:1994(en) Accuracy (trueness and precision) of measurement methods and results — Part 1: General principles and definitions*. Geneva, Switzerland: International Organisation for Standardization, 2002.

33. Arnold Kaufmann and Madan M. Gupta. *Introduction To Fuzzy Arithmetic: Theory And Applications*. New York: Van Nostrand Reinhold, 1985.

34. Dieter Klaua. "Ein Ansatz zur mehrwertigen Mengenlehre". In: *Mathematische Nachrichten* 33.5-6 (1967), pp. 273–296.

35. Jacob Klein. *Greek Mathematical Thought and the Origin of Algebra*. Translated by Eva Brann. New York: Dover Publications, Inc., 1992.

36. George Jiří Klir and Bo Yuan. *Fuzzy Sets and Fuzzy Logic: Theory and Applications*. Upper Saddle River, New Jersey, 07458: Prentice Hall PTR, 1995.

37. A. Kolmogoroff. *germanGrundbegriffe der Wahrscheinlichkeitsrechnung*. Berlin: germanVerlag von Julius Springer, 1933.

38. David H. Krantz et al. *Foundations of Measurement Volume I: Additive and Polynomial Representations*. New York: Academic Press, 1971.

39. Rudolf Kruse et al. *Computational Intelligence: A Methodological Introduction*. 2nd edition. London: Springer-Verlag, 2016.

40. William Labov. *Sociolinguistic Patterns*. Seventh printing 1978. Philadelphia: University of Pennsylvania Press, 1972.

41. George Lakoff. "Hedges: A study in meaning criteria and the logic of fuzzy concepts". In: *Journal of Philosophical Logic* 2.4 (1973), pp. 458–508.

42. Jeong-Gon Lee and Kul Hur. "Bipolar Fuzzy Relations". In: *Mathematics* 7.11 (2019), article number 1044.

43. Adrian Mehic. "Student beauty and grades under in-person and remote teaching". In: *Economics Letters* 219 (2022), p. 110782.

44. Günter Menges and Heinz J. Skala. "On the Problem of Vagueness in the Social Sciences". In: *Information, Inference and Decision*. Ed. by Günter Menges. Dordrecht: Springer Netherlands, 1974, pp. 51–61.

45. Asesh Kumar Mukherjee et al. "A Brief Analysis and Interpretation on Arithmetic Operations of Fuzzy Numbers". In: *Results in Control and Optimization* 13 (2023), p. 100312.

46. Giuseppe Peano. *Selected works of Giuseppe Peano translated from the Italian*. Translated and edited, with a biographical sketch and bibliography, by Hubert Collings Kennedy. Toronto: University of Toronto Press, 1973.

47. Xindong Peng and Ganeshsree Selvachandran. "Pythagorean fuzzy set: state of the art and future directions". In: *Artificial Intelligence Review* 52.3 (2019), pp. 1873–1927.

48. Surapati Pramanik. "Single-Valued Neutrosophic Set: An Overview". In: *Transdisciplinarity*. Ed. by Nima Rezaei. Cham, Switzerland: Springer International Publishing, 2022, pp. 563–608.

49. Willard Van Orman Quine. "Introduction to Moses Schönfinkel's 1924 "On the building blocks of mathematical logic" ". In: *From Frege to Gödel: A Source Book in Mathematical Logic, 1879–1931*. Ed. by Jean van Heijenoort. Cambridge, Massachusetts: Harvard University Press, 1967, pp. 355–357.

50. Magdaléna RenÄ?ová. "An Example of Applications of Intuitionistic Fuzzy Sets to Sociometry". In: *Cybernetics And Information Technologies* 9.2 (2009), pp. 43–45.

51. Roland Sambuc. "Fonctions -floues: Application a l'aide au diagnostic en pathologie thyroidienne". PhD thesis. Université de Marseille, France, 1975.

52. Berthold Schweizer. "1 - Triangular norms, looking back—triangle functions, looking ahead". In: *Logical, Algebraic, Analytic and Probabilistic Aspects of Triangular Norms*. Ed. by Erich Peter Klement and Radko Mesiar. Amsterdam: Elsevier Science B.V., 2005, pp. 3–15.

53. Florentin Smarandache. "Indeterminacy in Neutrosophic Theories and their Applications". In: *International Journal of Neutrosophic Science* 15.2 (2021), pp. 89–97.

54. Florentin Smarandache. "Neutrosophic Set—A Generalization of the Intuitionistic Fuzzy Set". In: *2006 IEEE International Conference on Granular Computing*. May 2006, pp. 38–42.

55. Nicholas J. J. Smith. *Vagueness and Degrees of Truth*. New York: Oxford University Press, 2008.

56. Michael Smithson and Jay Verkuilen. *Fuzzy Set Theory: Applications in the Social Sciences*. Quantitative Applications in the Social Sciences 147. Thousand Oaks, California: Sage Publications, Inc., 2006.

57. Oswald Spengler. *The decline of the West*. An abridged edition by Helmut Werner. English abridged edition prepared by Arthur Helps from the translation by Charles Francis Atkinson. Oxford, UK: Oxford University Press, 1991.

58. T. Sudha and G. Jayalalitha. "Fuzzy triangular numbers in - Sierpinski triangle and right angle triangle". In: *Journal of Physics: Conference Series* 1597.1 (2020), p. 012022.

59. Apostolos Syropoulos. "3 Vague theory of computation". In: ed. by Apostolos Syropoulos and Basil K. Papadopoulos. Berlin, Boston: De Gruyter, 2021, pp. 33–44.

60. Apostolos Syropoulos. "Fuzzifying P Systems". In: *The Computer Journal* 49.5 (July 2006), pp. 619–628.

61. Apostolos Syropoulos. "On Generalized Fuzzy Multisets And Their Use In Computation". In: *Iranian Journal of Fuzzy Systems* 9.2 (2012), pp. 113–125.

62. Apostolos Syropoulos. "Should We Take Vagueness Seriously?" In: *Philosophy Now* 156 (2023), pp. 36–39.

63. Apostolos Syropoulos and Theophanes Grammenos. *A Modern Introduction to Fuzzy Mathematics*. New York: John Wiley and Sons Ltd, 2020.

64. Apostolos Syropoulos and Basil K. Papadopoulos, eds. *Vagueness in the Exact Sciences: Impacts in Mathematics, Physics, Chemistry, Biology, Medicine, Engineering and Computing*. Berlin, Boston: De Gruyter, 2021.

65. Apostolos Syropoulos and Eleni Tatsiou. "2 Vague mathematics". In: ed. by Apostolos Syropoulos and Basil K. Papadopoulos. Berlin, Boston: De Gruyter, 2021, pp. 19–32.

66. Arturo Tozzi. "Bipolar reasoning in feedback pathways". In: *Biosystems* 215–216 (2022), p. 104652.

67. I.B. Turksen. "Measurement of Fuzziness: An Interpretation of the Axioms of Measurement". In: *IFAC Proceedings Volumes* 16.13 (1983). IFAC Symposium on Fuzzy Information, Knowledge Representation and Decision Analysis, Marseille, France, 19-21 July, 1983, pp. 97–102.

68. UNDP (United Nations Development Programme). "2023 Global Multidimensional Poverty Index (MPI)". In: *UNDP (United Nations Development Programme)* (2023).

69. Javad Vahidi and Salim Rezvani. "Arithmetic Operations on Trapezoidal Fuzzy Numbers". In: *Journal of Nonlinear Analysis and Application* (2013). The publisher is a defunct publishing company and the paper is available from https://www.researchgate.net/publication/263441550_Arithmetic_Operations_on_Trapezoidal_Fuzzy_Numbers.

70. Jay Verkuilen. "Assigning Membership in a Fuzzy Set Analysis". In: *Sociological Methods & Research* 33.4 (2005), pp. 462–496.

71. William Voxman. "Canonical representations of discrete fuzzy numbers". In: *Fuzzy Sets and Systems* 118.3 (2001), pp. 457–466.

72. Nigel Wheatley. "A sorites paradox in the conventional definition of amount of substance". In: *Metrologia* 48.3 (2011), pp. L17–L21.

73. Ronald R. Yager. "On The Measure Of Fuzziness And Negation Part I: Membership In The Unit Interval". In: *International Journal of General Systems* 5.4 (1979), pp. 221–229.

74. Ronald R. Yager. "On the theory of bags". In: *Int. J. General Systems* 13 (1986), pp. 23–37.

75. Ronald R. Yager. "Pythagorean Membership Grades in Multicriteria Decision Making". In: *IEEE Transactions on Fuzzy Systems* 22.4 (2014), pp. 958–965.

76. Shunsuke Yatabe and Hiroyuki Inaoka. "On Evans's Vague Object from Set Theoretic Viewpoint". In: *Journal of Philosophical Logic* 35.4 (2006), pp. 423–434.

77. Lotfi Aliasker Zadeh. "Fuzzy Sets". In: *Information and Control* 8 (1965), pp. 338–353.

78. Lotfi Askar Zadeh. "A Fuzzy-Set-Theoretic Interpretation of Linguistic Hedges". In: *Journal of Cybernetics* 2.3 (1972), pp. 4–34.

79. Lotfi Askar Zadeh. "The concept of a linguistic variable and its application to approximate reasoning—I". In: *Information Sciences* 8.3 (1975), pp. 199–249.

80. Lotfi Askar Zadeh. "The concept of a linguistic variable and its application to approximate reasoning—II". In: *Information Sciences* 8.3 (1975), pp. 301–357.

81. Lotfi Askar Zadeh. "The concept of a linguistic variable and its application to approximate reasoning—III". In: *Information Sciences* 9.1 (1975), pp. 43–80.
82. Wen-Ran Zhang. "Bipolar fuzzy sets and relations: a computational framework for cognitive modeling and multiagent decision analysis". In: *NAFIPS/IFIS/NASA '94. Proceedings of the First International Joint Conference of The North American Fuzzy Information Processing Society Biannual Conference. The Industrial Fuzzy Control and Intellige.* 1994, pp. 305–309.
83. Hans-Jürgen Zimmermann. *Fuzzy Set Theory—and Its Applications.* 3rd ed. Boston: Kluwer Academic Publishers, 1996.

Index